"REAL"

Medical Finals

OSCE

Survival Guide

Written by

Dr Ricky Frazer

Series Editor

John Frazer

Second edition 2013

ISBN: 978-1-4710-6865-2

Contents page

Clinical examinations

Practical Procedures

History Taking - *page 142*

Acute Management - *page 155*

Communication in a difficult or challenging situation - *page 165*

Practice mark sheet - *page 177*

Mark sheet - *page 178*

Preface

It is with immense excitement that I present to you the second edition of my second book. It is several years since I started writing this book and in its conception I envisioned a text that combined all the best bits I have learnt as a medical student and a junior doctor. The drive for the book came simply from my own experience revising for the final year OSCE and I hope with all sincerity that you enjoy reading it as much as I did writing it.

This book would not have been possible without the invaluable enthusiasm and unwavering support of the people close to me. I would like to thank my father and brother John for all their time and effort in editing and designing this book and for tolerating my unrealistic demands when it came to publication! I would like to express my gratitude to my mother Sharon who continues to inspire me every day. A special mention of my two beautiful nieces, Keria and Isabella. Also, a special thanks goes out to my sister Louise, who is a source of strength for everything I do. I would like to extend my appreciation over the pond to my friends in America and Canada, notably Courtney, Shannon, Kirk, Brenda, Vuella and Manon. I would also like to thank my teachers at Barry Comprehensive School, Brian Williams and Malcolm Phillips, who believed in my potential long before others saw any! Finally I extend a special thank you to Raj, a colleague and a close friend who contributed to many of the ideas for this book.

Ricky Frazer

January 2013

Introduction

So you have made it this far... well done! Many have failed along the way and you're still here with only the small matter of your final exams before that beautiful day in August where it all begins for real!

You have probably done enough online questions by now to feel confident that you can wing the single best answer paper, but OSCE are not so easy to wing!

By this point you have probably used a number of books to learn your clinical examinations and now it's a case of just learning one routine for each examination, so that when nerves take over on the day you can just perform the same examination you have practiced time and time again.

Having completed my finals in Cardiff 4 years ago I decided to pull together all the best bits I had picked up from wards, teachers and demonstrations in the lead up to my finals and incorporate it into one book. So here it is, this is the book that got myself and 4 friends through that dreaded exam and now I hope it will do the same for you.

The problem we faced when practicing together is that we had all used different books and had different steps for each examination that we learnt. Using the same system meant that practicing on each other and on patients became very straightforward and very easy to see what we were missing (e.g. point 8 of the abdominal examination, which I always seemed to miss!).

I must emphasise that this is not a book to explain the meaning of all the signs you elicit, there are many upon many examination books that will do this for you... rather this book gives you the steps that you need to perform and know off by heart to pass with flying colours!

I have concentrated most of this book on the core examinations and practical procedures. I have also included sections on acute management, history taking and communication skills. Look at the examples given and attempt (with a few colleagues) the scenarios provided.

The best way, in fact probably the only way to pass finals, is to examine each other and patients in groups. Take this book with you when seeing patients and use the self assessment table at the back to be critiqued by your fellow students. At the end of each day the progress you have made will be contained within this book. You can easily identify which examination you have practiced, how many times you have practiced it and which numbered points you forgot. This allows you to re-learn these areas until they are perfect!

I have provided causes, investigations and management for the common cases seen in the main systems; these include respiratory, abdominal, cardiology and neurology. For all clinical examinations I have provided a common cases section and Food for thought section to help develop your knowledge on areas that could be covered in the time left with the examiner after you have completed your clinical examination.

Finally, I am so confident that you *will* pass that this book comes with a guarantee. After using this book, if you don't pass, I will give your money back. Just log on to www.rickyfrazer.com

So that's it, you're ready to go. Don't forget that many of us have been there before. We passed and so can you... good luck!

Structure of the book

Each clinical examination has a section at the start that outlines *things to remember...* these are usually things that I forget to do in my examination or little tips to make it seem effortless. Each examination is then set out point by point so that it is easy to mark each other on the progress chart provided at the back to highlight which numbered points have been missed and need a little more work.

After the examination steps there is a section titled *for finals* which contains points relevant to the OSCE exam. I then present a list of *suitable cases for OSCE* and a *Food for thought* section which will give you some topics on which to quiz each other after performing the examination. For the core systems which are all very likely to feature in your OSCE exam in March, I have provided a section titled *Causes, Investigation and Management for common finals cases* which will give you an idea of the sort of questions the examiners WILL ask once you have finished your examination. Make sure you practice presenting these to each other.

Equipment required

Before you start seeing patients it is worth purchasing some basic equipment to take around the wards with you (trust me you will never find a tendon hammer on the ward when you need one). If you're feeling brave I suggest you dust off that old bum bag from the nineties and take your equipment around in that!

It is suggested that you have:
- Stethoscope
- Watch with a second hand
- Pen torch or ideally an ophthalmoscope
- Tendon hammer
- White and red hat pins
- Tuning fork
- Cotton wool
- Orange stick or keys (for testing the plantar reflex)
- Tape measure
- Reading chart to test visual acuity

Objective Structured Clinical Examination (OSCE) Structure

20th, 21st and 22nd of March 2013 in Out-patient Department, UHW and Clinical Skills Centre, Cochrane Building

There will be 15 examination stations covering clinical examinations, practical procedures, communication skills, history taking and the assessment and management of acute emergencies. The examination is divided into parts A and B.

Part A (Stations 1-10)

- 6 clinical examination stations
- 1 acute management station
- 1 communication station (difficult consultation)
- 2 focused history at stations

Eight and a half minutes per station with one and a half minutes change over time.

Part B (Stations 11-15)

- 4 practical procedure stations
- 1 communication station (explanation station)

Seven minutes per station with one minute change over time.

You will be assessed on 4 Practical Procedures which will be chosen from the list below:

- Arterial blood sampling
- Resuscitation
- Subcutaneous and intramuscular injections
- Use of peak flow meter/inhalers - (explanation/demonstration of procedure to the patient
- Intravenous cannulation
- Intravenous injections
- Catheterisation
- Rectal examination (on model)
- Suturing
- Phlebotomy
- ECG - placing of electrodes and interpretation of an ECG

All of these OSCE stations will be marked by a standardised mark sheet and you also have to achieve the pass mark in 10 of the 15 stations to pass the examination overall.

Good Luck!

Clinical Examinations

Abdominal examination

Things to remember

- A large number of signs in the abdominal examination can be found upon peripheral inspection
- In renal transplant patients, candidates can put together a storybook about previous renal replacement (AV fistula, haemodialysis line, peritoneal line), previous immunosuppressive treatments (steroids - cushinoid, ciclosporin - gum hypertrophy), function of the kidney (fluid status, loin tenderness, uraemic flap) and cause of renal impairment (finger prick - diabetes)
- Look for signs of chronic liver disease - if absent start to consider haematological causes and malignancy for hepatomegaly
- Other signs may favour a particular cause of chronic liver disease - Dupuytren's contracture, parotid swelling - alcohol, bronze and finger prick - haemachromatosis)
- Check to see if signs of decompensation are present - jaundice, ascites, encephalopathy and liver flap

ICE

1. Introduce
2. Consent
3. Expose - ask the patient if they can remove their top and lie flat and check if they are in any pain or discomfort

Inspection

4. Inspect from the end of the bed - general inspection, obvious jaundice, abdominal distension, renal transplant scar
5. Look at the hands - clubbing, leuconychia, koilonychia, palmar erythema, Dupuytren's contracture and test liver flap - check for AV Fistula
6. Look in the patients eyes - jaundice, anaemia, Keiser fisher rings, xanthelasma
7. Look around the mouth - pigmentation around lips, angular stomatitis, telangiectasia
8. Look inside the mouth - smooth enlarged tongue, aphthous ulcers
9. Smell the breath - fetor hepaticus, ketones
10. Feel for lymph nodes in the neck - especially left supraclavicular fossa
11. Check the back for spider naevi
12. Inspect for axillary hair and pigmentation
13. Look at the chest - gynaecomastia, spider naevi (if present must blanch one - to do this, press on the central blob and very quickly remove your finger - if quick enough you will see the blood rush back and red spider legs reappear),
14. Look for any obvious abdominal distension, periumbilical and flank bruising, scars
15. Lightly palpate the abdomen each of the 9 quadrants
16. Deep palpation of the 9 quadrants
17. Bimanually palpate the kidneys
18. Feel for the liver edge

19. Feel for the spleen
20. Feel for an aortic bruit
21. Feel for any inguinal hernia (1cm above the mid inguinal point), ask patient to cough
22. Percuss from the umbilicus out to each side, check for shifting dullness
23. Percuss the liver and spleen
24. Listen for bowel sounds
25. Listen for aortic bruits
26. Listen for renal bruits
27. Tell examiner that you would complete the examination by performing a PR
28. Thank the patient and cover them back up
29. Thank the examiner
30. Wash your hands
31. Present your findings

For finals

- Hepatomegaly – remember to percuss down the right side of the chest to rule out displacement of the liver by emphysema
- Splenomegaly – enlarges in to the left hypochondrium, massive splenomegaly can reach the RIF. It has a notch in its border, moves downward with inspiration and is dull to percussion
- Chronic liver disease – leuconychia, palmar erythema, Dupuytren's contracture, spider naevi, purpura, jaundice, loss of axillary hair, testicular atrophy, caput medusa, hepatomegaly, splenomegaly and ascites

- Ascites - abdominal distension, everted umbilicus (sometimes), fluid thrill, dull percussion in flanks and demonstrate shifting dullness
- Enlarged kidneys - palpable in the flanks using a bimanual technique, resonant to percussion
- When presenting a renal transplant patient: Try to present the findings using 4 headings: 1) Forms of renal replacement therapy (Haemodialysis line , PD catheter, AV fistula - are they being used?) 2) Evidence of immunosuppression (ciclosporin - gum hypertrophy, corticosteroids - cushinoid features 3) Function of the kidney (fluid status, loin tenderness, uraemic flap) 4) Cause of renal failure - diabetes (injection sites abdomen, finger pricks), APKD (bilateral masses in flanks), Alport's disease (hearing aid)

Suitable cases for OSCE

Common

Hepatomegaly	Liver failure & cirrhosis
Splenomegaly	Ascites
Hepatosplenomegaly	Pelvic kidney (transplant)
Polycystic kidneys	Ileostomy, colostomy

Less common

Hernias	Lymphadenopathy
Primary biliary cirrhosis	Mass RIF
AAA	IVC obstruction

Causes, Investigation and Management for common finals cases

Hepatomegaly

Causes - Can be remembered as MACCIVE HM (as in massive HepatoMegaly)

- Metabolic, autoimmune, cirrhosis, congestive cardiac failure, infections, hepatic vein thrombosis, ethanol, haematological, malignancy

Investigations

- Blood tests: FBC, clotting, urea and electrolytes, LFT and glucose
- Liver screening bloods
- Abdominal ultrasound scan
- Liver biopsy
- Tap ascites (if present)
- CT abdomen

Management

- Depends on underlying cause

Splenomegaly

Causes - can be divided up into massive, moderate and mild

Massive

- Haematological – CML and myelofibrosis

- Infectious – malaria, visceral leishmaniasis and kala azar

Moderate

- Sarcoidosis, amyloidosis and lymphoproliferative disorders

Mild

- Portal hypertension
- Infectious (infective endocarditis)
- Haemolytic anaemia

Investigation

- FBC, calcium, serum ACE, blood film, thick and thin films
- Viral serology
- CT chest and abdomen
- Lymph node biopsy
- Bone marrow

Management

- Depends on the underlying cause

Food for thought

What is the emergency management of haematemesis?
What are the causes of gynaecomastia?
What are the causes of chronic hepatitis?
How would you manage acute liver failure?
What is on your differential for ascites?
How would you manage ascites?

What is the management of acute ulcerative colitis?

What are the radiological differences between UC & Crohn's?

How would you investigate a patient with rectal bleeding?

Cardiovascular examination

Things to remember

- Offer to do the blood pressure
- Don't forget to look closely for pacemakers!
- It is often easier to see the cardiac impulse at the apex than feel it, so always inspect
- If the patient is young consider congenital problems such as ASD, VSD, coarctation and dextrocardia

ICE

1. Introduce
2. Consent
3. Expose – adjust the bed so that the patient is lying at 45° and ask patient if they can remove all of their clothes from their top half and roll up their trousers

Inspection

4. Stand at the end of the bed – perform a general inspection around the bed – oxygen, walking aids, do a toe to head inspection, look at legs for saphenous vein scar suggestive of a CABG, chest wall scars, pacemaker scars and cardiac facies (marfans)
5. Look at the hands – look at the back of hand first - clubbing, splinter haemorrhages, quinkes sign and tendon xanthomata. Ask patient to turn their hands over and look for Osler nodes, Janeway lesions, and rub the palms feeling for temperature

6. Feel for the pulse (do both at the same time so you can also check for radial-radial delay). Check rate, rhythm, volume and character (slow rising)

7. Ask patient if they have any pain in their shoulder - check for a collapsing pulse

8. Offer to take their blood pressure

9. Look around the eyes for any xanthelasma

10. Look in the eyes for signs of anaemia, corneal arcus

11. Look for any malar flush

12. Look around the mouth for angular stomatitis, dentition

13. Look inside the mouth for any cyanosis

14. Ask the patient to turn their head slightly to the left and look for an elevated JVP - measure from the sternal angle. Look both from the end of the bed and up close to the neck

15. Feel the carotid pulse (commenting on character)

16. Inspect their chest - look for any pacemaker, look for any sternotomy or thoracotomy scar - (if you see these, map them out with your finger to show the examiner you have both looked for them and found them). Look at the back for cardiothoracic surgery scars

17. Palpate for the apex beat and localise with one finger, map this out my counting down so to demonstrate to the examiner you know how to find it. Feel for the character of the apex - thrusting, heaving, double impulse

18. Palpate for any parasternal heaves (left of the sternum)

19. Palpate for any thrills over the aortic or pulmonary areas

20. Listen over the four areas with the diaphragm and with lighter touch (bell) over tricuspid and mitral valve while palpating the carotid pulse. Ask the patient to move on to the left side – repalpate the apex and listen again over mitral area and axilla in expiration to potentiate mitral murmurs

21. Ask the patient to sit forward and listen over the left sternal edge for aortic regurgitation in expiration

22. Listen over the carotid for carotid bruits

23. Listen to the base of the lungs for crackles

24. Feel for sacral oedema

25. Lie the patient back down and feel for peripheral oedema

26. If time allows, examine for hepatomegaly or a pulsatile liver

27. Say that to complete the examination you would examine the peripheral pulses, dipstick the urine for microscopic haematuria and check the blood pressure

28. Thank the patient and cover them back up

29. Thank the examiner

30. Wash your hands

31. Present your findings

Suitable cases for OSCE

Atrial Fibrillation
Mitral incompetence
Prosthetic valve
Mixed mitral valve disease

Mixed aortic valve disease
Left ventricular failure
Tricuspid incompetence

Less common

Raised JVP
Mitral stenosis
Eisenmenger's syndrome
Atrial septal defect

Dextrocardia
Complete heart block
Pulmonary hypertension
Coarcation of the aorta

For finals

Characteristics of the murmurs

- Mitral stenosis - pulse irregularly irregular, malar flush, cardiac impulse tapping, loud first heart sound, mid diastolic murmur heard, opening snap
- Aortic stenosis - pulse character slow rising, sustained, heaving, slightly displaced cardiac impulse, ejection systolic murmur heard all over the precordium and radiating into the carotids
- Aortic regurgitation - collapsing pulse, thrusting, displaced cardiac impulse, early diastolic
- Mitral regurgitation – displaced, thrusting cardiac impulse, soft S1, pan systolic murmur
- Aortic sclerosis – ejection systolic murmur heard all over the precordium but little radiation into the carotids

The best students will be able to make a comment about the severity of the murmur. Nobody is expecting you to act as an ECHO machine! ....but if you can grade severity based on clinical findings between "mild/moderate" and "moderate to severe" you really will PASS with flying colours.

Severity – moderate/severe characterised by:

Aortic stenosis - soft S2, narrow pulse pressure, slow rising pulse

Aortic regurgitation - collapsing pulse, widened pulse pressure, pulmonary oedema

Mitral regurgitation - signs of left ventricular failure (pulmonary oedema, tachycardia, gallop rhythm)

Mitral stenosis - increased duration of the murmur (just hearing a mitral murmur in itself is probably impressive enough!)

Causes, Investigation and Management for common finals cases

Aortic stenosis

Causes – can remember as ABCD
- Age related degeneration and calcification
- Bicuspid Valve
- Congenital (valvular, supravalvular - I'm sure you remember Williams syndrome in the deepest darkest part of your memory)
- Destruction (Rheumatic fever or endocarditis)

Investigation

CXR - valve sometimes calcified
ECG - left ventricular hypertrophy - don't forget your criteria
ECHO - looking at LV function, valve gradient and area
Angiogram - often coronary disease co exists. You may therefore consider performing the CABG and valve replacement at the same time

Management

The most important determinant of treatment (despite what some books might tell you) is symptoms. If the patient is symptomatic with chest pain, presyncope, syncope or breathlessness this is a strong indication for surgery.

Asymptomatic - no treatment or review in clinic of symptoms with ECHO evaluation

Symptomatic - balloon valvuloplasty (rare nowadays), aortic valve replacement or trancutaneous aortic valve implantation in patients that are high risk for surgery

Aortic regurgitation

Causes - ABCDE can be considered as those affecting the aortic valve and those that affect the aortic root

Ankylosing spondylitis/aortitis
Bicuspid valve
Connective tissue disorder (RA)

Dissection/drugs - pergolide
Endocarditis/Ehlos-Danlos

Investigations

ECG - lateral T wave inversion (representing strain)
CXR - cardiomegaly, mediastinal widening
ECHO - assess ejection fraction and aortic root
Angiogram - assess coronary arteries patency

Management

Medical - review of symptoms with echo assessment
Surgery - replace aortic valve when patient symptomatic

Mitral stenosis

Causes - rheumatic fever (most common), endocarditis, age related degeneration

Investigations

ECG - bifid P waves (P mitrale) and atrial fibrillation
CXR - left atrial enlargement, valvular calcification, pulmonary oedema
ECHO - assess valve area and flow, evidence of calcification or thrombus formation

Management

Medical - diuretics, warfarin + beta blockers/digoxin (if AF)
Surgery - valvuloplasty or valve replacement

Mitral Regurgitation

Causes - can remember as ABCDEF

A – Acute rupture/amyloid
B – Beta haemolytic streptococci (rheumatic fever)
C – Connective tissue/calcification
D – Drugs – pergolide
E – Endocarditis
F – Functional (left ventricular dilatation)

Investigation

ECG – Atrial fibrillation, P mitrale
CXR - enlarged heart, left atrial enlargement, pulmonary oedema (same as mitral stenosis)
ECHO – MR jet, valve area, dilatation of ventricle, ejection fraction, identify cause

Management

Medical – treat AF, diuretic, ACE inhibitors
Surgical – valve repair or valve replacement

Food for thought

What are the causes of atrial fibrillation?
How do we treat atrial fibrillation?
What are the side effects of amiodarone therapy?
What are the causes of mitral regurgitation?
What are the complications of mitral regurgitation?
What prosthetic valves do you know of?

What are the causes of aortic stenosis?

How do we investigate suspected AS?

What drugs do you know of which can exacerbate heart failure?

What drugs do we use to treat heart failure?

How do we manage acute heart failure?

What is the initial management of a patient with an MI?

What are the contraindications to thrombolysis?

What are the risk factors for ischemic heart pain?

Respiratory examination

Things to remember

- The respiratory exam is one of the hardest to fit in to 6 minutes, if still struggling with time omit TVF
- Horner's syndrome is characterised by miosis, partial ptosis and anhidrosis

ICE

1. Introduce
2. Consent
3. Expose - sit patient at 45° angle and expose the patient appropriately - check if they have pain anywhere

Inspection

4. General examination from the end of the bed
5. Look around for breathing aids – look for oxygen and a sputum pot (look inside it)
6. Check if comfortable/dyspnoeic at rest, breathing room air/on oxygen
7. Chest movements (ask patient to take a couple of deep breaths in and out - symmetrical chest movement, use of accessory muscles/pursed lips)
8. Ask the patient to lift up their neck – check to see if the trachea is central. Look from the side to see if there is an increased AP diameter
9. Take the respiratory rate (must remember to do!)

10. Look at the patient's hands - clubbing, tar staining, muscle wasting, peripheral cyanosis and tremor. Feel temperature - check the hands are warm and well perfused

11. Check for a flapping tremor

12. Take the pulse - rate, rhythm and volume

13. Inspect face for swelling (superior vena obstruction due to a lung tumour), eyes for Horner's syndrome, in eyes for anaemia

14. Ask the patient to poke out their tongue and then lift it up - looking for central cyanosis

15. Assess the JVP (right internal jugular vein) - measure if elevated, check for hepatojugular reflux

16. Palpate trachea (warn patient may be uncomfortable) and assess tracheal tug (sign of COPD - palpate the trachea during inspiration to see if it moves downwards)

17. Palpate the precordium for the apex beat position

18. Have a closer inspection of the neck and chest - (surgical emphysema) - look in both axilla for chest drain scars

19. Examine the neck for the presence of enlarged lymph nodes – perform from back

20. Ask the patient to sit forward and place their hands on their knees/cross hands to facilitate examination of the neck and back of the chest - check they are comfortable

21. Inspect the back of the chest for abnormalities of chest shape (kyphoscoliosis), other visible abnormalities (scars - look very closely for these, chest drains, bandage over previous chest drain scars). Check chest movement (ask patient to take a deep breath in and out through mouth - to see if movement is symmetrical)

22. Ask if the back of their chest is tender - if so ask the patient if it is ok to examine the area and gently palpate the painful area

23. Palpate for chest expansion posteriorly, comparing both sides - do at apex and base. Ask the patient to take a deep breath in, and then out and hold it, you should then place your hands on the back and ask them to take a deep breath in

24. Assess tactile vocal fremitus over the posterior aspect of the chest, comparing the two sides

25. Percuss the back of the chest, comparing the two sides

26. Auscultate over the back of the chest, comparing the two sides

27. Assess vocal resonance over the back of the chest and in the axillae, comparing the two sides

28. Check for sacral oedema

29. Ask patient to take a deep breath in and out and look for good and equal chest expansion - palpate chest expansion anteriorly (place hands very tightly around the patient) comparing both sides

30. Assess tactile vocal fremitus at the front of the chest and in the axillae comparing the two sides

31. Percuss over the front of the chest and the axillae comparing the two sides (when percussing the axillae ask the patient to place their hands behind their head)

32. Auscultate over the front of the chest and in the axillae (asking patient to breathe deeply through the mouth (comparing the two sides)

33. Assess vocal resonance at the front of the chest and in the axillae, comparing the two sides

34. Listen over the pulmonary area for the presence of a loud second heart sound (suggestive of pulmonary hypertension)

35. Check for peripheral oedema

36. Tell the examiner to complete the examination you would auscultate the heart and perform peak flow. State you would

also like to have a look at the patient's oxygen saturations and temperature chart
37. Thank the patient
38. Wash your hands
39. Present your findings

For finals

- Key signs that will often be found in finals are clubbing, reduced expansion (unilaterally or bilaterally) and abnormalities of percussion or auscultation especially at the bases posteriorly
- Pulmonary fibrosis - clubbing, fine crackles at bases of lungs bilaterally posteriorly
- Bronchiectasis - clubbing, coarse crackles at bases of lungs bilaterally posterior/wheeze
- COPD - hyper expanded chest, percussion note resonant/hyper resonant, expiratory wheeze, decreased breath sounds
- Common causes of clubbing and bilateral basal crackles are - pulmonary fibrosis and bronchiectasis

Suitable cases for OSCE

Common

Pleural effusion
COPD
Emphysema
Fibrosing alveolitis
Consolidation of lung

Atelectasis
Bronchiectasis
Cor pulmonale
Cystic fibrosis
Lung Cancer

Less common

Old TB   Pancoast's syndrome
Systemic sclerosis   Pneumonectomy

Causes, Investigation and Management for common finals cases

Pulmonary Fibrosis

Causes - can be remembered as upper lobe – BREAST and lower lobe – SCARD

BREAST - berylliosis, radiation, extrinsic allergic alveolitis, ankylosing spondylitis, sarcoidosis, TB

SCARD - scleroderma/SLE, cryptogenic fibrosing alveolitis, asbestosis, rheumatoid, drugs (amiodarone/nitrofurantoin)

Investigations:

Bloods - determine cause – auto-antibodies (rheumatoid factor, anti DsDNA, ANA) Serum ACE
CXR – reticular nodular shadowing
ABG – type 1 respiratory failure
Pulmonary function tests - restrictive picture (FEV1/FVC more than 70%), reduced lung capacity, reduced transfer factor
High resolution CT scan - cystic appearance, reticular pattern

Management:

Treat underlying cause
Pulmonary rehabilitation
Stop smoking
Immunosuppression – steroids and steroid sparing agents (azothioprine)
N-acetyl cysteine (free radical scavenger)
LTOT
Lung transplantation

Bronchiectasis

Causes - can remember as the 5 C's (if you remember the two I's as well that's a bonus!)

Cystic fibrosis, Childhood infections (measles, pertussis, TB), Carcinoma (and other obstructive lesions), Connective tissue disorders (rheumatoid), Congenital (Kartageners, Young's syndrome), Immunodeficiency (hypogammaglobulinaemia) Immunooveractivity (ABPA)

Investigations:

Divided into those to identify the cause and general investigations

Causes - sweat chloride and genetic analysis, bronchoscopy, auto antibodies, immunoglobulins, saccharin test for mucocillary clearance

General

Sputm MCS and cytology - common pseudomonas colonisation
CXR - tramline and ring shadows
Pulmonary function test - mixed restrictive and obstructive picture
CT Chest -"signet ring appearance" - dilated bronchi with thickened bronchial wall

Management

Specialist physiotherapy
Mucolytics - carbocysteine
Bronchodilators
Prompt treatment of infective exacerbations (using previous sensitivities)
Prophylactic antibiotics in selected patients
Surgery - occasionally used for localised disease

Chronic obstructive pulmonary disease

Causes - smoking, industrial exposure, alpha 1 antitrypsin

Investigation

Bloods: WCC (infection) alpha 1 antitrypsin
ABG: Type 1 or type 2 respiratory picture
Pulmonary Function tests: (obstructive) $FEV1/FVC < 0.7$, FEV1 marker of severity, increased residual volume, decreased transfer factor

CXR - hyper expansion, narrowed mediastinum r/o infection/pneumothorax

Management

Smoking cessation
Pulmonary rehabilitation
Long term oxygen therapy
Bronchodilators
Vaccinations
Nutritional support
Surgical
Treat acute exacerbations - controlled oxygen, bronchodilators, steroids, antibiotics and consider NIV

Food for thought

What are the respiratory causes of clubbing?
What is an exacerbation of COPD?
How do we differentiate between restrictive and obstructive lung disease?
How is cystic fibrosis inherited? What problems do these patients get?
What are the causes of pleural effusion?
What investigations can we carry out to find the causes of a pleural effusion?
What are the indications for a chest drain in someone presenting with an effusion?
How would you investigate a patient with SOB and a pleural effusion?
What is the clinical presentation of sarcoidosis?
How would you investigate a patient with pleuritic chest pain?

Neurological examination of the arm - motor

Things to remember

- Be prepared to combine the motor and sensory examinations
- Examine for pronator drift
- There are different characters of tone - hypotonia - due to either LMN or cerebellar lesions. Spastic rigidity - due to UMN lesions. Lead pipe/cogwheel rigidity - parkinsonism

ICE

1. Introduce
2. Consent
3. Expose - take the patient's shirt off and check for pain

Inspection

4. Look around the bed - check for any walking aids, foot splints. Look at the patient's face and chest for any neurological stigmata (café au lait)
5. Ask the patient to put their hands out in front of them, palms down with eyes closed - look for resting/postural tremor/athetosis
6. Ask the patient to put their hands out in front of them, with palms up and eyes closed - look for pyramidal drift or upward movement (cerebellar)
7. Inspect the muscles and arms for any wasting, scars or fasciculations (tell the examiner that you would normally wait

for a full minute). If no fasciculations are seen you can try to elicit them by tapping

Tone

8. Ask the patient to relax their body and muscles. Passively flex and extend the wrists and elbow as well as pronating and supinating the forearm. Note the presence of increased or reduced tone. To accentuate increased tone ask them to move the other hand up and down

Power

9. Each muscle group should be assessed in isolation. You should test each arm symmetrically comparing the grade of power. Shoulder abduction - (deltoid - C5) - ask the patient to "raise your elbows like a chicken. Don't let me push them down"

10. Shoulder adduction – ask the patient to "push down" against you (from the chicken position)

11. Elbow flexion (biceps - C6) – "put your arms up like a boxer. Don't let me pull your arms out" – support the elbow with one hand

12. Elbow extension (triceps - C7) "push me out"

13. Wrist flexion (C7-C8) – "cock your wrists down – don't let me push them up" (remember "like with like" use your wrist against theirs)

14. Long wrist extensors (C6-C7) - "cock your wrist up, don't let me push it down"

15. "Cock you wrists down – don't let me push them up"

16. Finger extension (extensor digitorum - C7 radial) - "extend your fingers rigid like a board and stop me from pushing them down"

17. Finger flexion (grip - C8) - "clasp my fingers and squeeze them as hard as possible"

18. Thumb abduction (abductor pollicis brevis – T1) "raise your thumb to the ceiling don't let me push it down" (checks median nerve – remember LOAF muscles)

19. Finger abduction (1st dorsal interosseus – T1) – "spread your fingers apart – don't let me push them in" (checks ulnar nerve)

Reflexes

20. Place each arm across the chest in turn. Use a tendon hammer to elicit the reflexes and compare both sides (biceps reflex - C5, 6). If reflex absent use the reinforcement technique by asking the patient to clench their teeth

21. Supinator – test for the supinator reflex by hitting the radial aspect of the forearm just above their wrist (supinator reflex C5, 6)

22. Triceps – bend the elbow to 90° and observe the muscle contraction (triceps reflex - C7)

Co-ordination

23. Perform the finger - nose test by asking the patient to "touch your nose and then touch my finger as fast as you can but as accurately as you can". Look for an intention tremor and past pointing (dysmetria). Ask them to do it with their eyes open and then move back and fore to their nose with

their eyes closed, as it will dramatically worsen if the patient has a sensory ataxia

24. Test for dysdiadochokinesia. Make sure the patient's hand comes off the other hand each time it moves up and down. Again ask them to perform "as fast as you can but as accurately as you can"

25. Tell the examiner that to complete the examination you would do a full sensory examination and examine the lower limbs and cranial nerves

26. Thank the patient

27. Wash your hands

28. Present your findings

Neurological examination of the arm - sensory

ICE

1. Introduce
2. Consent
3. Expose - ask the patient to take their shirt off and ask if they have any pain

Inspection

4. Look around the bed looking for any walking aids, foot splints, look at the patient face and chest for any neurological stigmata
5. Ask the patient to put their hands out in front of them with their palms down and eyes closed - look for resting/postural tremor/athetosis
6. Ask the patient to put their hands out in front of them, with palms up and eyes closed - look for pyramidal drift or upward movement (cerebellar)
7. Inspect the muscles and arms for any wasting, scars or fasciculations (tell examiner you would normally wait a full minute) - if there are no fasciculations seen you can try to elicit them by tapping

Pain - pin prick (spinothalamic)

8. Request to test pain sensation using a neuro pin. Firstly reassure the patient that you are going to test pain sensation and that it should feel sharp but wont puncture the skin. Apply the pin to the sternum asking "does it feel sharp like

you would expect a pin to feel?" Move proximal to distal mapping the dermatomes. Compare dermatomes in each arm

9. Start distally and move up the arm in a non dermatomal pattern looking for any distal peripheral sensory neuropathy. Ask the patient to tell you if the feeling changes as you move up the arm. Compare both sides

Light touch (dorsal column)

10. Place cotton wool over the sternum and ask "does it feel soft like cotton wool?" Ask the patient to close their eyes, so they are unable to obtain any visual clues, and ask them to respond verbally to each touch

11. Put the patient's arms in the anatomical position (palms facing upwards). Apply cotton wool to the dermatomes within the arms. Compare both sides symmetrically

Proprioception (dorsal column)

12. Ask the patient to watch what you are doing. Hold the distal interphalangeal joint of the index finger by its sides with your thumb and forefinger of one hand and move up and down telling the patient "this is up, this is down". Next, perform the same movements with the patients eyes closed and ask them to identify if you moved it up or down. Move proximally to larger joints if they cannot identify the movements

Vibration sense (dorsal column)

13. Place the tuning fork over the sternum and check they can feel the vibration. Place it over the interphalangeal joint of

the thumb and ask them if they can feel it vibrating and to tell you when it stops. Move proximally to larger joints if they cannot identify the vibration

14. Tell the examiner that to complete the examination you would like to perform a full motor exam and examine the lower limbs and cranial nerves

15. Thank the patient

16. Wash your hands

17. Present the findings

For finals

- Test like with like so check finger abduction by resisting with your finger

Suitable cases for OSCE

Wasted hand
Ulnar nerve palsy
Median nerve palsy
Radial nerve palsy
C5, C6 root
Upper motor neurone lesions

Hemiplegia
Parkinsonism
Peripheral neuropathy
Multiple sclerosis
Proximal myopathy

Less common

Nystagmus
Spastic paraparesis

Pseudobulbar palsy
Cerebellar signs

Causes, Investigation and Management for common finals cases

Hemiplegia/hemiparesis

Causes
CVA
Tumour
Trauma
Multiple sclerosis

Investigation

Bloods - FBC, fasting glucose, fasting Lipids, consider ESR/thrombophillia screen in young patient's
ECG - atrial fibrillation
CXR - aspiration
CT Head - infarction or haemorrhage
Carotid dopplers/ECHO/24 hour tape - embolic source

Management - (acute and chronic management – remember multidisciplinary)

Acute
Thrombolysis (within 4.5 hours of symptoms and if not contraindicated), Aspirin 300mg 2/52, SALT assessment (NBM), hydration, thromboprophylaxis, MDT input (stroke team), cautious with BP reduction

Chronic
Address secondary risk factors, arfarin after 2/52, carotid endarterectomy, rehabilitation, psychological/nutrition

Food for thought

How would you investigate a patient with UMN signs on physical examination?

What are the causes of UMN lesions (unilateral/bilateral)?

What are the signs of upper motor neurone lesions?

What are the causes of a wasted hand?

What would you expect in T1 root palsy? What are the causes?

What is the difference on clinical examination of a T1 root and ulnar nerve lesion?

What are the causes of median nerve palsy?

Can you name the muscles in the hand supplied by the median nerve?

What are the causes of LMN signs in the arm?

Can you name some causes of proximal myopathy?

Neurological examination lower limb - motor

Things to remember

- Always compare both sides
- For an absent reflex use the reinforcement technique by asking the patient to interlock their fingers and tighten them
- You can assess gait at the start or the end – but don't forget it

ICE

1. Introduce
2. Consent
3. Expose - adequately expose the lower limbs. Before beginning the examination, ask the patient if they are in pain

Inspection

4. Perform a general inspection of the patient and around the bed – look for any walking aids and perform a general surveillance of the patient from head to toe. Look in particular for evidence of pes cavus in the feet
5. Perform a more detailed examination of the limbs - look for any obvious scars, muscle wasting or fasciculations

Tone

6. Test tone - ask the patient if they have any pain in their back or legs. Ask them to relax their muscles. Roll each leg from side to side and lift the knee off the bed whilst looking for movement at the feet (heel should rub along bed - if

flicks up it suggests increased tone). Check for clonus by externally rotating the leg and slightly bending the knee and quickly dorsiflexing the foot – 1 – 2 beats can be normal

Power

7. Compare each leg. Hip flexion (iliopsoas – L1, L2) – ask the patient to "lift your leg up and don't let me push it down"
8. Hip extension (gluteus max – S1) - place your hand under their calf and tell the patient "push down into my hand"
9. Knee flexion (hamstrings - L5, S1) – "lift your leg and bend your knee, now pull your heel towards your bottom" – support their knee with your other hand
10. Knee extensors (quadriceps L3, L4) – from flexed position "push your leg towards the ceiling"
11. Ankle dorsiflexion (tibialis anterior L4) – "bring your toes up don't let me push them down"
12. Ankle plantar flexion (gastrocnemius and soleus S1) – "push your toes down, don't let me push them up"
13. Big toe extension (extensor hallucis longus - L5) – "lift your big toe up, don't let me push it down"

Reflexes

14. Rest the patient's knee flexed on your arm. Rub the spot on the knee that you're going to strike (knee reflex - L3, L4)
15. Ankle reflex – roll the leg out to the side and bend it slightly. Place the foot in dorsiflexion. When hitting the achilles tendon feel for plantar flexion of the foot and look at the calf muscles for contraction (ankle reflex - S1, S2)
16. Plantar response – use an orange stick or keys and draw an L shape on the lateral aspect of the foot. A normal response

is a down-going plantar, an up-going plantar suggests an upper motor neurone lesion

Co-ordination

17. Heel/shin - ask the patient to run their heel from their knee down to their ankle along the shin. Ask them to repeat this action but to make an arc in the air when moving back from the ankle to the knee (slightly more complex task). Repeat the test on the other leg

Gait

18. There are four parts to the assessment of gait. Ask the patient to walk on tip toes and then on their heels (will potentiate foot drop). Ask them to "walk heel to toe as if you are walking on a tight rope". This will highlight truncal ataxia. Ask patient to walk normally looking for any common abnormalities of gait
20. Romberg's test - ask the patient to stand up tall with their eyes shut. Observe if the patient is less stable (positive Romberg's test) whilst doing this. Romberg's sign will be positive in sensory ataxia rather than cerebellar ataxia

Completion

21. For completion, state you would perform a full sensory exam and then examine the upper limbs and cranial nerves
22. Thank the patient and offer to help them get dressed
23. Wash your hands
24. Present your findings

Neurological lower limb - sensory

ICE

1. Introduce
2. Consent
3. Expose - the lower limbs. Ask the patient if they are in pain

Inspection

4. Perform a general inspection of the patient and look around the bed – look for any walking aids and perform a general surveillance of the patient from head to toe – look in particular for pes cavus in the feet

Gait

5. Remember there are four parts to gait assessment. Ask the patient to walk to the end of the room, turn around and return. Look for any common abnormalities of gait and arm swing. Ask them to walk on tiptoes and then on their heels (will potentiate foot drop). Ask them to "walk heel to toe as if walking on a tightrope" – this will highlight truncal ataxia
6. Romberg's test – ask the patient to stand up tall with their eyes shut. Observe if the patient is less stable (positive Romberg's test). Romberg's sign will be positive in sensory ataxia

Pain - pin prick (spinothalamic)

7. Request to test pain sensation using a neuro pin. Firstly reassure the patient that you are going to test pain sensation

and that it should feel sharp but will not puncture the skin. Apply the pin to the sternum and ask "does it feel sharp_like you would expect a pin to feel?" Move from proximal to distal down the leg mapping out the dermatomes. Compare the dermatomes in each leg

Light touch (dorsal column)

8. Place cotton over the sternum and ask "does it feel soft like cotton wool?" Ask the patient to close their eyes, so they are unable to obtain any visual clues, and ask them to respond verbally to each touch
9. Apply cotton wool to the dermatomes of the legs. Compare both sides symmetrically
10. Start distally and move up the leg in a non dermatomal pattern looking for any distal peripheral sensory neuropathy. Ask the patient to tell you if the feeling changes as you move up the leg. Compare both sides

Proprioception (dorsal column)

11. Ask the patient to watch what you are doing. Hold the distal interphalangeal joint of the big toe by its sides with your thumb and forefinger of one hand and move it up and down telling the patient "this is up, this is down". Next, perform the same movements with the patients eyes closed and ask them to identify if you moved it up or down. Move proximally to larger joints if they cannot identify the movements

Vibration sense (dorsal column)

12. Place the tuning fork over the sternum and check that they can feel the vibration. Place it over the interphalangeal joint of the big toe and ask them if they can feel it vibrating (not just the cold of the tuning fork) and to tell you when it stops. Move proximally to larger joints if they cannot identify the vibration

Completion

13. For completion, state you would perform a full motor exam and then examine the upper limbs and cranial nerves
14. Thank the patient and offer to help them get dressed
15. Wash your hands
16. Present your findings

For Finals

- Examine for dermatomal and distal sensory loss

Suitable cases for OSCE

Hemiplegia   Peripheral neuropathy
Motor neurone disease   Parkinsonism
Charcot Marie tooth disease   Spastic   paraparesis

Causes, Investigation and Management for common finals cases

Peripheral neuropathy

Causes - can remember as ABCDE or ABCDEFGH if feeling clever!
Alcohol
B12 deficiency
Carcinoma
Diabetes, Drugs (isoniazid/vincristine)
Every vasculitis
Familial (Charcot Marie Tooth)
Guillian Barre
Hypothyrodism/HIV/Hepatitis

Investigations
Drug and alcohol history
FBC (MCV), B12, glucose, autoimmune screen, ESR
Nerve conduction studies

Management
Depends on underlying cause
Most common cause is diabetes
Management of diabetes - multidisciplinary, good diabetic control, manage micro vascular and macro vascular complications, advice regarding foot care

Food for thought

What are the causes of hemiplegia as seen in this patient?

What do you understand by the term upper motor neurone disease?

What are the causes of spastic paraparesis?

How would you investigate this patient with peripheral neuropathy?

What are the causes of sensory peripheral neuropathy?

What is foot drop and what causes do you know of?

What is a hemiplegic gait?

Can you give me some causes of cerebellar disease?

What is a syringomyelia? What are the signs?

What is syringobulbia and what are the additional signs to syringomyelia?

What are the neurological complications of chronic alcohol disease?

Cranial nerve examination

Things to remember

- If asked to examine the fundi, but find that you cannot get an adequate view, check if the pupils are constricted and tell the examiner that you are unable to properly visualise the fundus - it may be that the eye drops have worn off!

ICE

1. Introduce
2. Consent
3. Expose - shouldn't be needed for this one!

Inspection

4. Look for facial asymmetry, ptosis

Systematic examination

5. Olfactory nerve - ask the patient if they have noticed a change in their sense of smell
6. Optic nerve – check the visual acuity using a Snellen chart. If a Snellen chart is not available use words on a newspaper or a book. If the letters can not be seen, progress to finger counting, and then to finger movements if acuity is very poor
7. Check pupil responses. Ask the patient to fixate on an object in the distance. Inspect the pupil size, shape and check the direct and consensual light reflex with a pen torch

8. Check for a rapid afferent papillary defect (RAPD). Perform the swinging light test - swing a light from one eye to the next, observing for sustained pupillary constriction. Interrupted constriction suggests a relative afferent pupillary defect (common cause – multiple sclerosis)

9. Offer to perform fundoscopy with particular emphasis on the red reflex, visualisation of the optic disc, the four quadrants and the macula

10. Accommodation - ask the patient to fixate on an object in the distance and then to look at a finger held close to the patients face. Observe for any changes in the size of the pupil

11. Check if they can see the whole of your face or if any part is missing

12. Assess visual fields by asking the patient to cover one eye with you covering the opposite eye and check when they can see the red hat pin coming in from the peripheries diagonally (can be performed with just wiggling fingers). Next establish the blind spot and check for the presence of a central scotoma using a red hat pin (ideally with the red hat pin attached to a stick). Check when the head of the pin appears and disappears superiorly, inferiorly and lateral in both directions

13. Check voluntary movement saccades. Ask the patient to keep their head still and put your hands either side of head. As the patient to look at your nose, then look at your clicking fingers then back at your nose then to your clicking fingers on your other hand.

14. Visual inattention – test by moving wiggling fingers either side of the patients head and ask them to report if they saw one or both (a parietal lobe lesion will cause only the ipisilateral wiggling fingers to be observed)

15. Occulomotor/trochlear/abducens - inspect for ptosis (3rd nerve palsy, Horner's), pupil size, strabismus (squint) or proptosis. Slow pursuit - fix the patient's head and ask the patient to follow your finger with their eyes. Perform from 1 meter away to avoid convergence and draw an H shape and then move up and down in the midline (this checks pursuit movement). Ask the patient if they have any double vision or pain at any time. If patient has double vision elicit whether the images are separated vertically or horizontally and in which direction the separation is maximal. Close one eye and note which image disappears (outer or inner image – outer image disappears when the eye of the problem is closed)

16. Trigeminal nerve - sensation - ask the patient if they have noticed any numbness or altered sensation (pain) in the face. Use cotton wool to test light touch in the ophthalmic, maxillary, and mandibular dermatomes. Compare the right and left areas, asking if both sensations were equal. Offer to repeat this with a pin tip

17. Motor - inspect the muscles of mastication for wasting and test strength by asking the patient to carry out the following commands - clench their teeth (masseters /temporalis), open mouth against resistance (pterygoids), move the open jaw from side to side (masseters)

18. Corneal reflex and jaw jerk - offer to perform this, state that you would test this with a wisp of cotton wool on the cornea

19. Facial nerve - inspect facial asymmetry and asymmetrical wrinkling of the forehead (Bell's palsy). Tell the patient you are going to check the strength of the muscles in the face. Test the muscles of facial expression by asking the patient to carry out the following commands - "raise eyebrows right up, close your eyes as tight as you can, don't let me open them,

close your mouth as tightly as you can and don't let me open it up"

20. Vestibulocochlear nerve - hearing - ask the patient if they have noticed any changes in their hearing. Cover one ear and rustle your finger next to the other and ask if they can hear it. Repeat the other side

21. Offer to perform Weber's and Rinne's test using a 512Hz tuning fork

22. Glossopharyngeal and vagus nerves - gag reflex - indicate that you would test the gag reflex by using an orange stick to touch the pharynx. More simply ask the patient to say "ahh" and look for ulnar deviation using a pen torch and tongue depressor

23. Hypoglossal nerve – ask the patient to open their mouth and rest their tongue observing for wasting and fasciculation (LMN). Ask the patient to stick out their tongue observing for deviation of the tongue (towards the side of the lesion). Ask the patient to move the tongue from side to side

24. Accessory nerve – ask the patient to shrug their shoulders (trapezius) against resistance and then "look over your shoulder and push into my hand" – feel the belly of the sternocleidomastoid with your other hand whilst performing this

25. Say to complete my examination I would like to examine the upper and lower limbs

26. Thank the patient

27. Wash your hands

28. Present your findings

For finals

- You need to know at least the basics of how to hold a fundoscope!
- If you do get fundoscopy (and I'm sure you hope that you don't) it will likely be one of diabetic retinopathy or hypertensive retinopathy - know the differences

Suitable cases for OSCE

Facial nerve palsy	Ptosis
Bulbar palsy	Third nerve palsy
Horner's syndrome	Myasthenia gravis

Causes, Investigation and Management for common finals cases

Facial nerve palsy

Causes - (think anatomical from pons to parotid)
Pons - multiple sclerosis, stroke
Cerebellar-pontine angle - tumour (acoustic neuroma)
Auditory canal - Bells palsy (most common), cholesteatoma
Face - parotid trauma, tumour
Mononeuritis multiplex - diabetes, vasculitis, sarcoid, amyloid, Lyme disease

Investigation

Bloods - glucose, ESR, serum ACE, Lyme serology
CT/MRI head - neoplastic lesion

Management

Bells palsy - reassurance, prednisolone (if it is early in the presentation), antiviral, eye protection

Food for thought

What are the causes of ptosis and how can we differentiate between them?

What is Horner's syndrome and give me some causes?

Do you know of any causes for a unilateral III nerve palsy?

What are the causes of a large pupil?

What are the causes of cataracts?

What systemic conditions can present with eye signs?

What is a bulbar palsy and name some conditions leading to this?

What is a pseudobulbar palsy and give some examples?

Hip examination

Things to remember

· In a patient with kyphosis, look specifically for an adduction deformity
· In a patient with lordosis you might expect to see a flexion deformity
· Perform the Trendelenburg test
· Measure actual and apparent limb lengths

ICE

1. Introduce
2. Consent
3. Expose - ideally the patient should undress to their underwear

Look

4. Ask the patient if they would mind standing for you
5. Observe their stance from the front looking for any obvious wasting of the quadriceps muscle or scars, check the alignment of the shoulders, hips and patella. Ensure the ASIS's are aligned and at the same level as each other
6. Look from the side for any deformity, swelling, redness or lumbar lordosis
7. Look from the back for any signs of wasting of the gluteus muscles or for any scoliosis
8. Ask the patient if they could take a few steps forward for you, turn around and walk back (observe the gait)

9. Perform the Trendelenburg test - used to assess hip stability. Ask the patient to stand on each leg in turn, resting their outstretched hands in yours, and lift the other leg off the ground by bending it at the knee. The sign is positive if the pelvis drops on the side that is unsupported

10. Lay the patient flat on the couch and inspect the hip region more closely looking for swelling, deformity or erythema

11. Measure apparent and actual limb lengths - square the pelvis making sure the iliac crests are in the same plane before measuring the length of the limbs. True leg length is from the ASIS to the medial malleolus. Apparent leg length is the distance from the xiphisternum to the medial malleolus. True shortening may be caused by avascular necrosis or hip dislocation. False shortening may be caused by an adduction deformity of the hip

Feel

12. Check if the patient has any pain and then palpate over the hips comparing the temperature

13. Feel for any swelling over the hips (effusions)

14. Feel for tenderness over the greater trochanter (trochanteric bursitis)

15. Feel for tenderness of the ischial tuberosity (hamstring tear)

Move

16. Ask them to flex and extend their hip keeping their leg straight then perform the same movements passively to the maximum range of movement

17. Ask them to flex and extend their hip with their knee flexed. Perform the same movements passively to the maximum range of movement

18. Fix one hip stabilising the pelvis by placing your hand on the contralateral ASIS. With your hand on the pelvis, first check active, then passive abduction

19. Ask the patient to adduct their hip and then perform passively (remember to stabilise contralateral ASIS)

20. Place leg at 90°- check for internal and external rotation by placing one hand on the knee and one hand on the ankle

Special tests

21. Perform the Thomas test – place your hand in the small of the back to ensure that the lumbar lordosis has been eliminated. Flex both hips then ask the patient to keep one flexed whilst straightening the other. If the limb is elevated off the examining table and cannot be straightened fully there is a fixed flexion deformity of the hip on the affected side. The angle through which the thigh is raised from the couch is the angle of fixed flexion. The test should then be repeated for the other hip

22. Tell the examiner that to complete the examination you would perform a neurovascular examination and formally examine the joints above and below (back and knee)

23. Thank the patient and offer to help them dress

24. Wash your hands

25 Present your findings

For finals

- Osteoathritis - can be primary or secondary. On examination you may demonstrate a positive Trendelenburg's sign. It is important to be aware of the typical radiographic changes and basic treatment approaches

Suitable cases for OSCE

Osteoarthritis	Total hip replacement
Rheumatoid arthritis	Hip arthrodesis
Slipped upper femoral epiphysis	Antalgic gait

Food for thought

What is your differential for hip pain?
Where is 'hip pain' usually referred to?
What features of the history would make you suspect OA?
How would you investigate suspected OA of the hip?
What would you expect to see on a plain radiograph?

Breast examination

Things to remember

- Examine the tail of Spence

ICE

1. Introduce
2. Consent – advise the examiner and patient that it is normal procedure to have a chaperone present
3. Expose - ask the patient to remove her clothes to the waist and sit on the end of the bed with her hands relaxed by her side

Inspection

4. Perform a general inspection – look in particular for asymmetry of the breasts, deformity, swellings, scars and skin discoloration
5. Inspect the nipple for any inversion, discharge or any skin changes (peau d'orange (breast carcinoma), eczematous change of Paget's disease)
6. Ask the patient to place their hands on their head. Look for any tethering, indentation or asymmetry
7. Ask the patient to firmly press on their hips
8. Make it clear you are also inspecting the axillae for visible lymph nodes or presence of scars (axillary clearance)

Palpation

9. Ask the patient to lay on the examination couch at 45°

10. Ask the patient if they have any pain in either breast. If any pain is present start by examining the normal side (as always)

11. Ask the patient to place their hand behind their head, on the side of the breast that you're going to examine

12. Palpate in all four quadrants of the breast. Use the palmar aspect of your fingers. If the patient has larger breasts you can use the other hand to steady the breast tissue

13. Use a gentle rotatory movement to gently compress the breast tissue against the chest wall and feel for the presence of any masses

14. Examine in a concentric circular pattern moving gradually outwards from the nipple to the border of the breast

15. Use your thumb and forefinger to palpate the tail of Spence

16. If a swelling is present, note its site, size, shape, consistency, tenderness, mobility and temperature

17. Ask the patient if they have had any discharge from the nipple and explain that you are going to examine the nipple

18. Feel gently over the nipple area and gently attempt to express the nipple for any discharge

19. Rest the patient's right elbow in your right hand and abduct the arm to expose the axillae. Using your left hand palpate the groups of lymph nodes - anterior, medial, posterior and apex

20. Exam the left axillae this time by abducting and supporting the left arm with your left hand and using your right hand to palpate

21. Palpate the cervical and supraclavicular lymph nodes from behind the patient

22. Tell the examiner that to complete the examination you would palpate the liver, auscultate the lung bases and palpate the spine looking for evidence of metastasis
23. Thank the patient - offer to cover them up
24. Wash your hands
25. Present your findings

For finals

- It is important to fully expose the patient to ensure adequate inspection is performed
- Remember to exam the normal side first as with all examinations

Suitable cases for OSCE

Benign breast changes	Fibroadenoma
Fibrocystic disease	Carcinoma

Food for thought

How do we investigate breast lumps?
What is the triple assessment?
What forms of imaging are available to investigate breast lumps?
Which 2 views are used in mammographic screening?
What is the most common reason for a 30 year old presenting with a breast lump?
What is the most common reason for a 50 year old presenting with a breast lump?
What are the main types of breast cancer?

Knee examination

Things to remember

- Examine for the presence of effusions using the patellar tap test and the bulge test
- Collateral ligaments - the medial and lateral ligaments can be assessed by applying valgus and varus force at the knee with the leg held under one arm. The medial ligament is tested by abducting the ankle while pushing the knee medially. The lateral ligament is tested by adducting the ankle while pushing the knee laterally.

ICE

1. Introduce
2. Consent
3. Expose - ask the patient if they can expose their legs leaving on their underpants or shorts

Look

4. Perform a general inspection from the front with the patient standing, look for alignment of the shoulders hips and patella. Check for any valgus or varus deformity and any wasting of the quadriceps
5. Look from the side for any fixed flexion deformity
6. Look from the back for a swelling in the popliteal fossa that may represent a Bakers cyst
7. Assess gait - ask the patient to walk a few steps forward and back- look for abnormalities of gait (antalgic gait)

8. Ask the patient to lie supine and inspect the knee closely for erythema, swelling, effusion, scars or misalignment of the patella

Feel

9. Ask the patient if there is any pain in either knee and examine the normal knee first
10. Compare skin temperature in each knee
11. Perform the patellar tap test and bulge test to detect the presence of an effusion
12. Feel along the joint line for tenderness with the knee flexed at approximately 90°
13. Palpate around the knee joint feeling over the lateral and collateral ligaments, tibial tuberosity and palpate the popliteal fossa for the presence of a Baker's cyst
14. Offer to perform the patella apprehension test

Move

15. Tell the patient that you are now going to test active movements of the knee and to tell you if this causes any discomfort
16. Test knee flexion by asking the patient to bring their ankle up towards their bottom
17. Test for knee extension by asking the patient to straighten their leg as far as possible
18. Tell the patient that you are now going to assess passive movements. Explain that you will do the movements and that they should let you know if it causes any pain

19. Attempt to flex the knee to 140° with one hand on the knee and the other hand on the ankle. Feel for crepitus of the knee joint

20. Maximally extend the leg to -10° while feeling for crepitus in the knee

Special tests

21. Test collateral ligaments
22. Test cruciate ligaments – anterior and posterior drawer test
23. Offer to perform Mcmurray's test
24. To complete the examination say you would like to perform a neurovascular examination of the leg and examine the joint above and below
25. Thank the patient and offer to help them get dressed
26. Wash your hands
27. Present your findings

For finals

- The most likely pathology for finals will be osteoarthritis as this is very common in the general population

Suitable cases for OSCE

Common

Osteoarthritis	Rheumatoid arthritis
Baker's (popliteal) cyst	Cruciate ligament tear

Less common

Subluxation of the patella   Meniscal tears

Food for thought

What are the causes of knee pain?
If you suspected OA, what investigations would you carry out?
What are the characteristic features on a plain radiograph?
What is a Baker's cyst?
What conditions it is associated with?
What is your differential for a baker's cyst?

Hand examination

Things to remember

- Froment's sign - ask the patient to clutch a piece of paper between their thumb and index finger. Attempt to pull the paper away from the clasp of the patient. If the thumb adductor is weak then the patient can only hold on to the paper by flexing the interphalangeal joints of the thumb and is unable to hold the thumb straight (ulnar nerve)
- Whilst examining the hands think about the different rheumatological conditions and look for their characteristic signs

ICE

1. Introduce
2. Consent
3. Expose - ask the patient to expose their arms to above their elbows. Place a pillow on the patient's lap and ask them to place their hands on it palm down. Ask the patient if they are in any pain and to tell you if you cause any discomfort

Look

4. Perform a general inspection of the patient - butterfly rash (SLE), beaked nose (scleroderma)
5. Look for nail changes - nail fold infarcts, clubbing, psoriatic changes - pitting, onycholysis, ridging

6. Look for the bony swellings of osteoarthritis (Heberden's nodes or Bouchard's nodes). Whilst inspecting the joints look for any acute inflammation

7. Actively look for rheumatoid changes, swollen joints, swan neck deformity, boutonniere deformity, z-shaped thumb, ulnar deviation and any classics signs of nerve palsy

8. Look for joint swellings with gouty tophi visible, check for tight skin (scleroderma)

9. Turn hands over and look at the palmar aspect. Look for palmar erythema, Dupuytren's contracture and look for any carpal tunnel scars (crease of the palm)

10. Look for muscle wasting – thenar eminence (median nerve), hypothenar (ulnar nerve) and the intrinsic muscles. Note any wasting of the long muscles of the arm

11. Look at the wrists for any swellings

Look elsewhere

12. Look at the elbows (ask patient to elevate their arms as if they are a boxer) look for psoriatic plaques, gouty tophi or rheumatoid nodules

13. Look in the helix of the ear for gouty tophi; inspect hair line for psoriatic plaques, the conjunctivae for episcleritis/scleritis (rheumatoid arthritis) and umbilicus (common site for psoriatic plaques)

Feel - move from distal to proximal

14. Ask the patient to place their hands back on the pillow palms down and feel each joint individually. Feel for the type of swelling

15. Ask the patient to turn their hands over and feel over the patient's forearm, hands and fingers to assess the temperature

16. Feel the muscle bulk over the thenar and hypothenar eminence

17. Check for tenderness in the anatomical snuff box and over the head of the ulna and radius

18. Feel the radial pulse bilaterally

19. Sensory divisions - check sensation over the lateral border of the ring finger (median nerve), medial border of the ring finger (ulnar) and the web space between the thumb and index finger (radial)

Move

21. Assess wrist movements - ask the patient to make the prayer sign and inverted prayer sign (with elbows up). Check radial and ulnar deviation as well as pronation and supination

22. Assess fingers - ask patient to spread their fingers (finger extension) and make a fist (finger flexion)

23. Assess finger abduction (ulnar nerve)

24. Assess thumb abduction - place hands flat with the palms facing upwards. Test thumb abduction by asking the patient to point their thumb up to the ceiling and not to let you push it down with your thumb (median nerve)

Assess function

25. Ask patient to carry out everyday tasks such as undoing buttons, holding a pen or a key

Special tests

26. Froment's sign (ulnar nerve) – offer to perform this if low on time
27. Tinel's sign (median nerve) – offer to perform this if low on time

Complete examination

28. Tell the examiner that to complete the examination you would perform a full neurovascular assessment and examine the other joints
29. Thank the patient
30. Wash your hands
31. Present your findings

For finals
- Move from distal to proximal
- Avoid shaking hands at the start as this may cause pain or discomfort

Suitable cases for OSCE

Rheumatoid hand   SLE
Psoriatic arthropathy   Osteoarthritis
Dupuytren's contracture   Ankylosing spondylitis
Systemic sclerosis   Gouty arthropathy

Less common

Paget's disease   Dermatomyositis

Food for thought

How would you investigate a patient with features of RA?

What are the radiological differences between OA and RA?

What is the pattern of distribution of RA?

Do you know of any systemic complications of RA?

Do you know of any diagnostic criteria for RA?

What other conditions is RA associated with?

What are the systemic signs of SLE?

Shoulder examination

Things to remember

ICE

1. Introduce
2. Consent
3. Expose - ask the patient to undress from the waist upwards

Look

4. Inspect the patient from the front, behind and sides. Look from the front for any obvious deformity, swelling or asymmetry. Inspect from the side for any wasting of the deltoid muscle, swelling, scars, redness or bruising. Look from the back for any muscle wasting or winging of the scapula

Feel

5. Ask the patient if they have any pain in their shoulders. Feel over the shoulder joint to assess temperature comparing each side
6. Palpate the bony landmarks of the shoulder for deformity, tenderness or effusions. Begin at the sternoclavicular joint, and then move along the clavicle towards the acromioclavicular joint (soft point between the clavicle and acromium)
7. Palpate below the clavicle for subacromial bursitis. Ask the patient to put their arms up like a boxer to contract the biceps muscle. Palpate the biceps tendon within the bicipital

groove (long head of the biceps). Pain on this manoeuvre may suggest biceps tendonitis

8. Palpate along the length of the spine of the scapula for any tenderness

Move (active then passive)

Active

9. Tell the patient that you are going to ask them to perform some movements at the shoulder and to inform you if they experience any pain. Note if there is limited range of movement and the degree of limitation

10. Abduction - ask the patient to raise both their hands sideways and put their hands together above their head (normal range 180°). If pain is present, establish the angle when it begins within the painful arc. Pain midway through the arc suggests supraspinator tendonitis or partial rotator cuff tear. Acromioclavicularjoint arthritis is suggested by pain towards the end of the arc.

11. Adduction - ask the patient to cross their arms across the front of their body (normal range 50°)

12. Flexion - ask the patient to raise their arms forwards (normal range 180°)

13. Extension - ask the patient to swing their arms backwards (normal range 60°)

14. Test for external and internal rotation (passively and actively with their arms by their side). Ask the patient to bend their arms and tuck their elbows into their side and separate their hands (normal range 45°). Ask the patient to perform the same movement but this time move their hands together (normal range 55°)

15. Test external rotation in abduction - ask the patient to put their hands behind their head

16. Test internal rotation with adduction by asking the patient to put their hands behind their back and try and touch their scapula

Power

17. Test deltoid power – ask the patient to put their arms up like a chicken and don't let you push their arms down

18. Test power of flexion and extension – ask the patient to "Put your arms up like a boxer. Don't let me pull your arms out". In the same position, ask the patient to "Push me out" (feel over the biceps and triceps whilst performing).

19. Test power of internal (subscapularis) and external rotation (infraspinatus and teres minor)

20. Perform Gerber's lift off test (subscapularis)

21. Perform Jobe's test (empty can test) - abduct arm to 90° - bring arms forward (scapula plane) and pronate (thumbs down). Tell the patient not to let you push their arms down (supraspinatus)

Passive – crepitus

22. Stand behind the patient, rest one hand on their shoulder and move their arm in all directions. Observe for crepitus, pain and limitation of movement

Special tests (if time allows)

23. Neer's (impingement) test

24. Hawkins test

25. Apprehension test
26. To complete the examination say that you would like to perform a neurovascular examination of the arm. Also say that you would like to perform a back and elbow examination
27. Thank the patient - offer to help them put their clothes back on
28. Wash your hands
29. Present your findings

For finals

- Impingement syndrome – affects the rotator cuff causing shoulder pain. It occurs due to repetitive movement - symptoms include pain (often worse at night) and weakness - gives a painful arc between 60 - 120° of abduction. Treatment - physiotherapy, steroid injections
- Rotator cuff tears - supraspinatus, infraspinatus and subscapularis tears are often due to chronic tendinitis (partial tears) or a sudden strain caused by a fall (complete tear). The findings on examination may include a painful arc in partial tears. Complete tears restrict shoulder abduction to just 60° and when lowering the arm below 90° the arm will suddenly drop (drop arm sign)

Suitable cases for OSCE

Frozen shoulder   Impingement syndrome
Rotator cuff tear   Winging of the scapula

Less common

Bicipital tendonitis   Osteoarthritis

Food for thought

What is adhesive capsulitis or 'frozen shoulder'?
What is impingement syndrome?
What are the causes of rotator cuff tears?
What is winging of the scapula?

Back examination

Things to remember

- Most likely diagnosis is ankylosing spondylitis
- Ensure that you perform a detailed inspection as this should give you the diagnosis in most cases

ICE

1. Introduce
2. Consent
3. Expose - ask the patient if they can undress to their underpants

Look

4. Ask the patient to stand up
5. Inspect from the back. Look at the skin for scars, pigmentation or abnormal hair
6. Look at the spine for changes in posture and any wasting or fasciculations over the paravertebral muscles
7. Inspect from the side looking for any changes in posture - increased thoracic kyphosis and loss of cervical and lumbar lordosis
8. Inspect from the front for any obvious asymmetry - look at the alignment of the chest, trunk and pelvis
9. Tell the patient that you would first like to see them walk. Ask them if they normally use a walking aid and if so ask them to use it. Assess the speed, phases of walking (heel strike, stance, push off and swing), stride length, arm swing and any known abnormal gaits (e.g. ataxic, high stepping, antalgic, festinant)

Feel

10. Palpate the full length of the spine over the cervical, thoracic and lumbar spine. Palpate over the paraspinal muscles to detect any tenderness

Move (can either tell them what to do or ask them to copy your movements)

11. Ask the patient to place their chin on their chest (cervical flexion)
12. Ask the patient to push head backwards (cervical extension)
13. Ask patient to put their ear on their shoulder (lateral flexion)
14. Ask the patient to look back over each shoulder (cervical rotation)
15. Ask the patient to keep their legs straight and lean backwards as far as they can whilst ensuring they don't fall backwards (lumbar extension)
16. Ask the patient to touch their toes while keeping their legs straight (lumbar flexion)
17. Ask the patient to slide their hand down their leg as far as they can whilst keeping their leg straight (lumbar lateral flexion)
18. Ask the patient to sit down and place their hands on their hips to ensure the pelvis is stable. Ask the patient to twist from side to side (thoracic rotation)
19. Assess chest expansion. Ask patient to take a full expiration and then a full inspiration. Normal expansion should be more than 5cm

Special tests

20. Schober's test - (request a tape measure). Make a mark at the level of the dimples of Venus (level L5 - posterior superior iliac spines). Measure 10 cm above the point and hold the tape measure at this level. Ask the patient to bend forward and touch their toes and measure the distance from the point back to the dimples of Venus. The distance should be at least 15cm

21. Straight leg raising test - ask the patient to lie on their back. Slowly flex the hip whilst keeping the leg in extension. Raise the patient's leg off the couch until the patient experiences pain. The pain is commonly experienced in the thigh, buttock or back and suggests sciatica.

22. Ask the patient to lie on their front. Palpate the sacroiliac joints for tenderness

23. Femoral stretch test – with the patient still on their back, flex the knee and extend the hip by gently lifting the leg. Pain suggests ipsilateral irritation of L2, L3, and L4 root

Completion

24. State to the examiner that to complete the examination you would like to perform a neurovascular examination

25. State that you would also like to examine the hip and shoulder joint

26. Thank the patient and offer to help them put their clothes on

27. Wash your hands

28. Present your findings

For finals

- Know the basic treatment for ankylosing spondylitis
- Familiarise yourself with the systemic manifestations of ankylosing spondylitis

Suitable cases for OSCE

Ankylosing spondylitis   Osteoarthritis
Kyphosis   Prolapsed disc

Food for thought

What are the classic features of ankylosing spondylitis?
What else would be on your differential for someone with lumbar back pain?
What is meant by the term 'prolapsed disc'?
What would be your differential for a prolapsed disc?
How would you differentiate these on examination?
What investigations can we carry out to confirm prolapsed disc?

Inguinal hernia examination

Things to remember

- If the patient is frail, examine them in the position they are in
- If any lumps are present, describe them in terms of site, size, shape (circular or irregular), consistency, temperature, translumination, upper edge (able to palpate or not) and cough impulse
- Indirect hernias comprise approximately 80% and are more common in younger patients. Usually the hernia stops at the superficial ring
- Direct hernias comprise approximately 20%. If the hernia is reduced it is not controlled by occlusion of the deep ring (and cough impulse) and reappears through the superficial ring

ICE

1. Introduce
2. Consent - offer patient a chaperone
3. Expose - groin and external genitalia. Warm your hands before starting

Inspection

4. Look for abdominal or inguinal scars
5. Look for any masses in the abdomen, inguinal region and scrotum and ask the examiner if they would like you to examine the abdomen and the normal side first (they wont!)
6. Ask patient to sit up/bear down whilst inspecting femoral or inguinal region for hernia

Palpation

7. Ask the patient to give a forceful cough
8. Ask the patient if the mass is painful and palpate the mass determining its shape, size, position and borders. Place your finger over the lower border of the hernia and feel for a cough impulse and determine if it extends to the scrotum

Reduction

9. If there is a mass in the scrotum, ask the patient if they are able to reduce it – "can you move the mass back into the tummy?" If so, ask them to perform this (with the patient laying down)
10. If they are unable to reduce the mass you can offer to perform this yourself (gently). Try to pass it through the superficial and deep rings
11. Identify the ASIS
12. Find the pubic tubercle by sliding your hand up the leg and finding the first bony prominence
13. Display your anatomical knowledge of the structures by stating that the inguinal ligament is the line between the ASIS and the pubic tubercle and the deep ring is half way along this line and 1cm above
14. Place a finger over the deep ring with a finger above so that the space between the deep and superficial rings is not occluded. Ask the patient to cough. If there is a cough impulse in the deep ring but an absence of a lump in the superficial ring, it suggests an indirect hernia. If there is no cough impulse over the deep ring but a lump appears in the superficial ring it suggests a direct hernia

15. Now release your fingers over the deep inguinal ring and ask the patient to cough again. You should see if it re-enters the canal and moves past your fingers

16. If a hernia is present percuss the lump for a resonant percussion note (bowel involvement). If the mass is not resonant it is likely to just be omentum

17. If a hernia is present, listen over the lump for bowel sounds (bowel involvement). If no bowel sounds are present it is likely to just be omentum

18. Transilluminate the mass

Testicular examination

19. Ask the patient if there is any pain or tenderness in the testis

20. Inspection - inspect the anterior aspect of the scrotum. Inspect the posterior aspect of the scrotum by pulling on the posterior skin. Look for any skin changes or swellings

21. Palpation - ask the patient if there is any pain or tenderness in the testis (an acute tender enlarged testis is suggestive of torsion of the testis)

22. Roll the testicle between your thumb and index finger. Determine if both are palpable (absence of the testis would suggest orchidectomy or an undescended testis)

23. Locate the epididymis found above and posterior to the testis. Note any swellings

24. Feel along the spermatic cord found above the epididymis. Note any swellings (epididymal cyst)

25. Offer to transilluminate the testis to exclude a hydrocoele

Extension

26. Examine the inguinal lymph nodes. Feel along the inguinal ligament (anterior superior iliac spine to the tubercle) for the horizontal chain of nodes and over the medial thigh for the vertical chain of lymph nodes
27. Palpate the femoral arteries and auscultate for bruits
28. Stand up to see if there are any masses that become more visible (particularly on the normal side)
31. To complete the examination say you would examine the other side and inspect for a saphenofemoral varix (femoral aneurysm)
32. Thank the patient
33. Wash your hands
34. Present your findings

For finals

- A hernia is the protrusion of an organ or part of an organ through a deficiency in the wall of the cavity in which it is contained

Suitable cases for OSCE

Direct inguinal hernia   Indirect inguinal hernia

Food for thought

What is the mid-inguinal point?
What is the relevance of these two points?
What is the significance of the pubic tubercle?
How do we investigate hernias?

Vascular (arterial) examination

Things to remember

- Description of ulcers should include – site, size, shape, base, edge, depth and discharge
- Site - use the terms - anterior or posterior, medial or lateral to describe the site of any ulcer or swelling. Describe its position in relation to the nearest bony prominence.
- Size – ideally use a tape measure to measure both its width and length
- Shape – keep it simple! Describe the shape of the lump as circular, oval or irregular
- Base - note the colour (pink/yellow/white), penetration (tendon muscle/bone), tissue (granulation tissue/dead tissue), tumour (SSC)
- Edge – helps determine the etiology - flat sloping - venous ulcer. Punched out - indicates the rapid death of a whole thickness of skin (usually on insensible areas) in diabetes. Rolled edge - BCC, everted edge - SSC
- Depth - measure in millimetres
- Discharge - serous, sanguineous or purulent

ICE

1. Introduce
2. Consent – ask the patient if they are in any pain
3. Expose – ask the patient to undress to their underwear and lie on the couch

Inspection

4. Stand at the end of the bed and perform a general inspection of the legs. Start at the feet and move proximally. Inspect for dystrophic nail changes, amputated toes, gangrene, changes in skin colour, and loss of hair. Perform a detailed inspection for any ulceration. Look in between the toes and particularly over the main pressure points

5. If an ulcer is present describe it in terms of site, size, shape, base, edge, depth and discharge

6. Inspect the abdomen for any visible pulsations suggestive of an aortic aneurysm

Palpation

7. Warm your hands and feel to the left of the aorta to feel for any aortic aneurysms

8. Place your fingers either side of the aorta, see if any mass felt is expansile

9. Temperature - move distal to proximal using the back of your hand feeling the temperature of the soles of the feet and limbs. Note any change in temperature as you proceed

10. Capillary refill time - press the tip of the nail on both legs for 5 seconds and measure the time taken for the blanched area to turn pink after the pressure is released. The nail bed colour should take less than 3 seconds to return to normal

11. Repeat capillary refill time for hands

12. Pulses - feel for the presence of the radial pulse and brachial pulse and offer to take the blood pressure

13. Feel the femoral pulses on both sides - demonstrate to the examiner how you locate it (midpoint between the symphysis pubis and the ASIS)

14. Feel the popliteal pulses - ask the patient to bend their knee and place your thumbs on the tibial tuberosity, feel the pulse with all 8 finger tips

15. Feel the posterior tibial pulses - halfway between the medial malleolus and the prominence of the heel

16. Feel the dorsalis pedis pulses – lateral to the extensor of the big toe in the dorsum of the foot

Auscultation

17. Listen for bruits along the aorta, iliac, femoral and popliteal arteries

Special tests

18. Buerger's test - with the patients legs flat on the couch, elevate each leg slowly and look for the angle when it becomes pale (Buerger's angle). A patient without vascular disease should not experience colour change of the leg. Sit the patient up and ask them to hang their legs down over the side of the bed and observe the time taken for their legs to return to normal colour. The skin at first becomes blue (blood is deoxygenated in its passage through the ischaemic tissue), and then red (due to reactive hyperaemia)

Completion

19. To complete the examination, state that you would like to feel the carotid pulse and listen for a bruit, auscultate the heart and measure the ankle brachial pressure index

20. Thank the patient and offer to help them get dressed

21. Wash your hands

22. Present your findings

Vascular (venous) examination

Things to remember

- A scar over the long saphenous vein may suggest previous grafting for coronary artery bypass surgery
- Main causes of varicose veins - prolonged standing, fibroids, prior surgery or trauma, deep vein thrombosis, pregnancy, ovarian tumour, prolonged standing and familial

ICE

1. Introduce
2. Consent
3. Expose - perform the examination in the standing position

Inspection

4. Inspect from all sides. Look for generalised oedema of the lower limbs
5. Look for superficial pigmentation (haemosiderin deposition causes a dark speckled appearance)
6. Look for ulcers – non-healing ulcers especially around the medial malleoli
7. Inspect for lipodermatosclerosis – fibrosis of the skin due to chronic inflammation
8. Look for venous flare - fan shaped dilatation of superficial venules spreading from the ankle

9. Look for any previous scars - suggestive of previous vascular surgery (particularly look over the long saphenous vein)

10. Look for any obvious eczematous changes

11. Establish the location and distribution of any varicose veins. Look particularly along the long saphenous vein (groin to medial malleoli) and short saphenous vein (popliteal to lateral malleoli)

Palpation

12. Temperature - using the dorsum of the hand feel along the soles of the feet and up the patients legs. Note any changes in temperature

13. Palpate the skin of the lower leg looking and feeling for pitting oedema. Determine the upper limit of the swelling

14. Look for thrombophlebitis (hard inflamed and tender veins). Feel along the long and short saphenous vein for tenderness suggestive of phlebitis or hardness suggestive of thrombosis

15. Feel the saphenofemoral junction (4cm below and lateral to pubic tubercle for a saphena varix (saphenous vein dilatation as it joins the femoral vein). Test for saphenofemoral incompetence by asking the patient to cough at this point and feel for any impulse generated

Auscultation

16. Listen for bruits over a venous cluster which is suggestive of an arteriovenous fistula

Special tests

17. Trendelenburg's test – ask the patient to lay flat and elevate their leg to allow emptying of the superficial veins. Occlude the saphenofemoral junction with two fingers and ask the patient to stand. Remove your fingers. If the superficial veins refill, this indicates incompetence at the saphenofemoral junction

18. Tourniquet test - ask the patient to lay flat and elevate their leg to allow emptying of the superficial veins. Place a tourniquet tightly around the upper thigh and ask the patient to stand whilst observing below the tourniquet. Superficial veins filling below this level indicates incompetent perforators below the level of the tourniquet. Repeat the test down the leg until the veins below the tourniquet fail to fill. At this position the incompetent perforators lie above the level of the tourniquet

19. Tap test - place the finger of one hand at the bottom of a long varicose vein and tap above the site with the other hand noting the presence of an impulse (superficial incompetence)

Completion

20. To complete the examination offer to examine the peripheral pulses to assess arterial blood supply and perform an abdominal and pelvic exam (obstruction of the inferior vena cava). Request a Doppler ultrasound probe to listen to flow in the incompetent valves

21. Thank the patient and offer to help them get dressed

22. Wash your hands

23. Present your findings

For finals

- Venous ulcers are usually found medially in the gator region – arterial ulcers tend to be lateral or distal
- Remember a list of chronic venous changes and look specifically for these
- Know the signs of acute ischaemia = 6P's

Suitable cases for OSCE

Peripheral vascular disease   Varicose veins
Ulcers   Diabetic foot

Food for thought

What symptoms are varicose veins associated with?
What investigations do we carry out to confirm venous disease?
What are the management options in a patient with varicose veins?
What is a DVT and how does it present?
What investigations do we carry out to confirm a DVT?
How do we classify chronic limb ischaemia?
How do we manage chronic arterial disease of the lower limb?

Thyroid/Neck examination

Things to remember

- If asked to examine the neck - start in the neck, then complete the examination by examining for thyroid status
- To remember the eye signs in thyroid disease you can remember ROLE - retraction, ophthalmoplegia, lag, exophthalmos

ICE

1. Introduce
2. Consent
3. Expose - ask the patient to expose their neck

Inspection

Neck

4. Inspect the patient from the front and from the side. Ask the patient to elevate their chin slightly and look for scars (parathyroid/thyroid), nodes, masses or a visible goitre
5. Ask patient to stick their tongue out (if the lump moves on tongue protrusion it suggests a thyroglossal cyst - moves upwards in the midline)
6. Sip water – ask the patient to sip some water, hold it in their mouth and swallow whilst observing the lump (if the lump moves on swallowing it is of thyroid origin)
8. Inspect the oral cavity and throat using a pen torch for enlarged tonsils (infection or malignancy)

Palpation

9. Ask the patient if they are in any pain before starting palpation

10. Palpate the anterior and posterior lymph nodes (Pre-auricular, sub-mandibular, sub-mental, paratracheal, supraclavicular, deep and superficial cervical and occipital nodes)

11. Stand behind the patient and ask them to flex their neck slightly to allow for easier palpation

12. Place your hands on either side of the patient's neck and feel in the anterior and posterior triangles

13. Examine any mass by fixing it one side with one hand and palpating with the other. Feel the mass and determine its site, size, shape, consistency, mobility, tenderness, pulsatility and transillumination

14. Ask the patient to take another sip of water and hold it. Ask them to swallow it whilst you are palpating the lump

15. Feel for the tracheal position in the suprasternal notch. Note if the trachea is central or deviated

Percussion

16. Percuss down the midline of the neck (over the costal cartilages) and determine the lower limit of the thyroid. A dull percussion note is suggestive of retrosternal extension

Auscultation

17. Ask the patient to take a deep breath in and then out and hold it. Listen over the thyroid glanfor any bruits (thyrotoxicosis)

Thyroid status

18. Inspect the hands - look for acropachy (Grave's disease)
19. Feel the temperature of the skin
20. Ask the patient to put out their hands and place a piece of paper over the dorsum of the hands to look for any tremor
21. Feel the pulse for bradycardia or tachycardia
22. Look around the eyes for swelling of the conjunctiva (chemosis)
23. Look for any loss of the outer third of the eyebrow
24. Look for any lid retraction (white of sclera visible around the cornea)
25. Exophthalmos - stand behind the patient and lean over the top of them (Graves's disease)
26. Test for any ophthalmoplegia (test eye movements and ask the patient to report any double vision – Grave's disease
27. Test for the presence of lid lag – ask the patient to follow your finger up and down (eyelid lags behind the globe)

Special tests

28. Ask the patient if they have any pain in their legs - ask them to kneel on the chair and check the ankle reflex for slow relaxation (hypothyroidism)

Completion

29. Tell the examiner that to complete the examination you would like to auscultate the heart
30. Thank the patient
31. Wash your hands
32. Present your findings

For finals

- Practice describing the details of a lump. (Always start with the 3S's – site, size and shape)
- If the case is thyroid disease, the examiner is likely to ask you what questions you might ask the patient. If hyperthyroid ask about - heat intolerance, increased appetite, weight loss and diarrhoea. If hypothyroid ask about cold intolerance, decreased appetite, weight gain and constipation
- Be prepared to provide some basic investigations – thyroid function tests, autoantibody screen, ultrasound scan, uptake scan and fine needle aspiration

Suitable cases for OSCE

Common

Goitre   Lymphadenopathy
Thyroglossal cyst   Parotid swelling

Less common

Dermoid cyst   Carotid body tumour

Food for thought

What are the complications of Grave's disease?
What is de Quervain's thyroiditis?
What thyroid cancers do you know of?
What are the causes of generalised lymphadenopathy?

GALS screen

Things to remember

- Remember that GALS (gait, arms, legs and spine) is a screening test and that a detailed examination is not required
- It is often easier to show the patient what action you would like them to perform rather than tell them

ICE

1. Introduce
2. Consent
3. Expose - ask the patient to undress to their underwear

Inspection

4. Inspect the patient standing for any abnormal position or posture of the limbs (observe for symmetry between the shoulders, arms, hips and knees). Look for any swelling or deformity

Spine (Back)

Look

5. Inspect from the front (protuberant abdomen suggestive of lumbar lordosis)
6. Look from behind, look in particular for scoliosis

7. From the side - look for the normal cervical lordosis, thoracic kyphosis and lumbar lordosis (look for the classic question mark posture of ankylosing spondylitis)

Feel

8. Press on each vertebral body in turn. Try to elicit any tenderness

Move (can either tell them what to do or ask them to copy your movements)

9. Ask the patient to place their chin on their chest and then pull their head backwards
10. Ask the patient to put their ear on their shoulder and then look behind each shoulder
11. Ask the patient to touch their toes while keeping their legs straight and then straighten themselves back up
12. Check spinal rotation by asking the patient to turn their body to either side

Arm

Look

13. Perform a general inspection of the skin (psoriasis, rheumatoid nodules and muscle wasting)
14. Inspect the joints - look for any joint swellings. Inspect for rheumatological changes in the hands or changes of osteoarthritis (Heberden's and Bouchard's nodes)

Feel

15. Feel over the midpoint of each supraspinatus to elicit any tenderness
16. Joints - squeeze each hand at the level of the carpal and metacarpal joints and try to localise any tenderness or swellings

Move

17. Shoulders - check full external rotation and abduction (hands behind the head) and internal rotation (hands behind back and touch the scapula)
18. Elbows - check flexion and extension
19. Wrists - check flexion and extension. Ask the patient to make the prayer sign and inverted prayer sign
20. Hands - power grip. Test the strength of the grip by asking the patient to squeeze your finger
21. Precision pinch grip - test strength of the grip by trying to break the patients pinch

Legs

22. Ask the patient to lay on the couch

Look

23. Perform a general inspection of the skin and muscles
24. Inspect for any obvious joint swelling or deformity

Feel

25. Feel over the joints for any tenderness, warmth or swelling. Palpate each knee along the joint margin. Squeeze each foot and try to locate any tenderness

Move

26. With the patient supine, check for hip flexion and extension
27. Hold the knee and hip at 90° of flexion and internally and externally rotate the hip. Observe the patient's face as you do this to ensure that are not causing the patient unnecessary pain
28. Place one hand on the knee joint and flex and extend it, feel for any crepitus

Gait

29. Ask the patient if they normally use a walking aid
30. Ask the patient to walk. Observe their gait, stride length and any difficulty turning
31. Observe phases of gait - heel-strike, stance, push off and swing

Completion

32. To complete the examination say you would perform a more detailed joint examination
33. Thank the patient and offer to help the patient dress
34. Wash your hands
35. Present your findings

For finals

- The examiner may well ask - what three GALS screening questions would you like to ask? Suitable questions would be - (1) Do you have any pain or stiffness in your muscles, joints or back? (2) Can you dress yourself completely without difficulty? (3) Can you walk up and down the stairs without difficulty?

Suitable cases for OSCE

Ankylosing spondylitis   Rheumatoid arthritis

Food for thought

How do we investigate ankylosing spondylitis?
What are the complications of ankylosing spondylitis?
What are the management options available in ankylosing spondylitis?
How do we manage polymyalgia rheumatica?

Practical Procedures

Arterial blood gas sampling

Things to remember

- When explaining to the patient about the procedure, ask if they have had it done before. Warn that it can be a bit painful, but should be a quick procedure. Ask if they have any questions
- Check which is the patient's dominant hand and ideally take sample from other one
- Ensure that you tell the patient that it is likely to be a little painful!

ICE

1. Introduce
2. Consent - explain the procedure to the patient and explain why it is required. Confirm the patients name, date of birth and first line of their address before you proceed
3. Expose – confirm which the patient's non dominant arm is and expose that arm from below the elbow

Preparation

4. Wash your hands
5. Collect the equipment required - syringe and heparin (more commonly now a heparinised syringe) two arterial gas needles, alcohol swabs, cotton wool/gauze, tape, pair of non-sterile gloves and a sharps bin

Procedure

6. Inspect and palpate the radial artery on the non-dominant side
7. Put on the gloves
8. Clean the skin over the artery using an alcohol wipe and allow to dry
9. Take a 2ml syringe and draw up a small volume of heparin (0. l ml)
10. Attach the needle (blue)
11. Palpate the artery with the non-dominant hand and extend the wrist, ideally resting over a pillow
12. Hold the syringe in the dominant hand like you would hold a pen and fix the skin between the index finger and middle finger of your non dominant hand
13. Insert the needle through skin towards the artery
14. As you advance the needle, apply gentle traction on the plunger
15. On arterial puncture, allow the syringe to fill (normally fill with 2ml of blood)
16. Remove the needle from the artery
17. Apply pressure to the puncture site for 5 minutes - then inspect for the development of a haematoma
18. Discard the needle in the sharps bin
19. Expel any air from the syringe and apply the cap to the syringe
20. Label the syringe with the patient's details
21. Complete the request form with the patient details
22. Tell the examiner that you would immediately take the sample to the blood gas machine for analysis
23. Check the puncture site and apply a suitable dressing
24. Discard the gloves in the clinical waste bin

25. Thank the patient and check they are comfortable

Questions that you could be asked

- Which sites would you obtain a blood gas sample from?
- What precautions should be taken to prevent false readings?

Resuscitation

Things to remember

- This is one of the only times that ICE does not apply - instead think SSSS (safe, shout for help, shout, shake)

SSSS

1. Safe - before you assess the patient you need to ensure it is safe to approach
2. Shout for help - if anyone is present tell them to call an ambulance, give the location, and then to come back and bring an AED if available
3. Shout - assess the patent's conscious state by calling their name or asking if they are ok
4. Shake - assess the patent's conscious level by gently shaking their shoulders

Airway

5. Open the airway by gently tilting the head back and lifting the chin (head tilt, chin lift) - if a cervical spine injury is suspected then open the airway by jaw thrust only

Breathing

6. Check whether the patient is breathing. Put your ear next to the patient's mouth and look (for chest movement), listen (for breath sounds) and feel (for air on your cheek) for no more than 10 seconds

7. If not breathing and you are on your own you will need to call for an ambulance

Circulation

8. If not breathing start chest compressions
9. Identify the costal margin and the xiphisternum - place the heel of your hands two finger breadths above the xiphisternum and interlock your fingers
10. Position yourself vertically above the patient with your arms straight and shoulders above the wrists. Apply chest compressions – 5-6cms deep
11. Compression rate should be about 100 - 120 per minute

Combine chest compressions with rescue breaths

12. After 30 compressions reopen the airway
13. Pinch the patient's nose
14. Place your lips around the patient's mouth ensuring you have a good seal and blow steadily for 1 second. Check for chest rise and fall
15. Give a second rescue breath. Now return your hand to the chest and continue with 30 compressions
16. Continue at a rate of 30:2 until the patient shows signs of recovery or help arrives
17. Only recheck the patient if they show signs of life

Questions that you could be asked

- If you suspect cervical spine injury how could you maintain an airway?
- Where do you place your hands for chest compressions?
- At what rate should this be performed at?
- How long should you continue?

Subcutaneous and intramuscular injections

Things to remember

- Possible sites for intramuscular injections - mid deltoid - good for low volume injections such as those less than 5ml. It has the most rapid uptake of all intramuscular sites due to its good blood supply. Gluteal - upper outer quadrant of the buttock - excellent site for large volume injections. There is however a small risk of sciatic nerve and vessel injury. The rectus femoris which is the anterior lateral aspect of the thigh (vastus lateralis) is good for most injections especially depot injections and sedatives

ICE

1. Introduce yourself
2. Consent - explain the procedure to the patient and obtain consent. Check the identity of the patient
3. Expose - area to be injected with relation to anatomical landmarks explaining to patient why that site has been chosen

Preparation

4. Check that the packaging of all equipment is intact and is within the expiry date
5. Wash your hands and put on your gloves
6. Select the appropriate equipment - a 5ml syringe and 2 green 21G needles are usually appropriate (orange needle for

subcutaneous injections). Check you have an alcohol swab, gauze, plaster and sharps bin available

7. Consult the patient's drug chart to ascertain the correct drug, dose, diluent (if needed), mode of administration time and date for the drug to be given. Check for any allergies

8. Confirm the name of the drug on the ampoule as well as the strength and expiry date with a nurse or doctor

9. Prepare the drug to the appropriate dosage if required

10. Tap the syringe facing vertically upwards to collect any air at the top. Expel any air from within the syringe ensuring no drug is lost

11. Discard the needle in the sharps bin and attach a new needle to the syringe for administration

Procedure - intramuscular injection

12. Clean the chosen site with an alcohol wipe

13. Allow to dry

14. Stretch the skin around the chosen site and inform the patient that they will feel a sharp scratch

15. Hold the needle at 90° to the skin and introduce it leaving a third of the shaft of the needle exposed

16. Drawback – pull on the syringe to ensure that the needle has not penetrated a blood vessel. If no blood is aspirated, depress the plunger, inject the drug slowly and remove the needle. If blood appears, withdraw the needle completely, replace it and begin again at a different site. Explain to the patient what has happened

17. Apply compression to the injection site using cotton wool or gauze

18. Dispose of the needle and syringe in a sharps bin

19. Record in the appropriate documents that the injection has been given - record the time that it was administered and sign the prescription chart

20. Thank the patient and check that they are comfortable

Procedure - subcutaneous injection

12. Clean the chosen site with an alcohol wipe

13. Allow to dry

14. Push the skin between your thumb and index finger to raise the adipose tissue from the underlying muscle

15. Approach the skin at 45° and penetrate with a firm motion

16. Pull back on the plunger to ensure you have not entered a blood vessel

17. Inject the drug slowly looking for the formation of a bleb under the skin

18. Dispose of the needle and syringe in a sharps bin

19. Record in the appropriate documents that the injection has been given - record the time that it was administered and sign the prescription chart

20. Thank the patient and check they are comfortable

Questions that you could be asked

- Outline the possible sites that can be used for an intramuscular injection and outline advantages of each?
- What should be done before the medication is administered?
- What needle is used in elderly patients and why?
- How many needles are required to give the injection?

Use of peak flow meter - (explanation/demonstration of procedure to patient)

Things to remember

- Ensure the pointer on the peak flow meter is set to zero

ICE

1. Introduce
2. Consent - elicit patient's understanding of asthma and the role of peak flow measurement. Ask the patient if they are happy for you to demonstrate how this is performed
3. Expose – ask the patient to stand (if this is not possible, ask them to sit up straight)

Procedure

4. Place a fresh mouthpiece into the peak flow meter
5. Ensure the pointer is at zero on the numbered scale
6. Show the patient how to hold the meter horizontally with their fingers underneath as to ensure the marker is not obstructed
8. Instruct the patient to start by taking a breath in as deep as possible
9. Explain and demonstrate to the patient how to blow out as hard and as fast as possible into the peak flow meter
10. Read the value and write it down so that it is not forgotten
11. Tell the patient to repeat the test three times and take the best reading

12. Tell the patient to then record this on the peak flow chart provided and compare it to the previous results

13. Ask the patient if they have any questions

14. Ask the patient to repeat the steps involved

15. Check the patients reading against a standardised chart taking into account the age, height and sex of the patient

16. Ask the patient if they have any questions or concerns

17. Thank the patient

Questions that you could be asked

- How can we use peak flow values in grading the severity of asthma?
- What determines the normal ranges for peak flow values?

Use of inhalers - (explanation/demonstration of procedure to patient)

Things to remember

- At the end of the explanation station the examiner may ask you some simple questions. These may include - possible side effects of salbutamol - fast heart rate, shakiness or headaches. How often the salbutamol inhaler should be taken - two puffs PRN (as required) up to four times a day. When patient should seek medical advice - if they find they need their salbutamol inhaler more than 4 times a day or experience significant side effects

ICE

1. Introduce yourself
2. Consent - explain the importance of using inhalers properly and ask the patient if they are happy for you to demonstrate the technique to them
3. Expose - ask the patient if they are able to stand up (if not able to, ask them to sit up straight)

Procedure

4. Choose the correct (blue) inhaler and tell the patient that they should check that the inhaler is in date
5. Instruct the patient to shake the inhaler prior to use
6. Remove the cap from the mouthpiece
7. Hold the inhaler upright between their index finger and thumb
8. Tell them to hold their head in a neutral position

9. Ask them to take a deep breath out

10. Form a good seal around the mouth-piece with their lips

11. Depress canister at the start of the breath and take a deep breath in

12. Tell the patient to hold their breath for 10 seconds and then breathe out slowly

13. Check whether the patient understood the steps involved

14. Ask the patient to repeat the steps involved

15. Ask the patient if they have any questions or concerns

Questions that you could be asked

What is the difference between reliever and preventer therapies and which drugs are used for each?

What are the common colours of the inhalers and what do they contain?

What is the stepwise treatment of asthma?

Intravenous cannulation

Things to remember

- Commonly used cannula sizes - 16G grey, 18G green, 20G pink

ICE

1. Introduce
2. Consent - explain the procedure to the patient – "I have been asked to insert a thin plastic tube into a vein on the back of the hand. You are likely to feel a sharp scratch when the needle is inserted. Would that be ok?"
3. Expose - ask patient to roll up their sleeve or remove their clothing so that the antecubital fossa and dorsum of the hand are exposed

Preparation

4. Collect the required equipment - cannula, cotton wool, alcohol swab, saline flush, syringe, needle, tourniquet, gloves, cannula securer (tegaderm) and a sharps bin
5. Wash your hands
6. Support the limb and apply a tourniquet to the upper arm
7. Select and inspect an appropriate vein - dorsum of the hand preferably (avoid using antecubital fossa if possible). Ask patient to clench their fist in and out
8. Select an appropriate cannula size
9. Put on gloves

Procedure

10. Clean the patent's skin with an alcohol swab and allow the skin to dry

11. Remove the cannula from its packaging. Ensure that the needle moves freely in and out and remove the top

12. Anchor the vein by stretching the skin a few centimetres below the proposed insertion site

13. Warn the patient of a "sharp scratch". Insert the cannula at an angle of approxinately 30° and watch for the flash back of blood

14. Advance the cannula by a further 2mm into the vein

15. Advance the plastic cannula

16. Press your finger over the vein at the tip of the cannula

17. Remove the needle and place the cap on the cannula

18. Dispose of the needle into the sharps bin

19. Apply the tegaderm

20. Flush the cannula with a saline flush

21. Discard your waste and gloves into a clinical waste bin

22. Thank the patient and check they are comfortable

23. Wash your hands

24. Record the procedure in the patent's notes including the time and date

Questions that you could be asked

- What sites are recommended for cannulation?
- What sizes of cannula are available?
- What are the risks associated with intravenous cannulation?
- What is the need for cannulation?

Intravenous injections

Things to remember

- If it is a powder medication then add the correct diluent carefully down the wall of the ampoule, agitate the ampoule and inspect the contents
- Ensure the solution is clear before withdrawing the prescribed amount
- In the real hospital environment most intravenous injections are given into a cannula

ICE

1. Introduce - tell the patient who you are and confirm their name, date of birth and the first line of their address
2. Consent - explain the procedure to the patient – "I have been asked to give you a medication directly into the vein in your arm. You may feel a sharp scratch when the needle passes into the skin. Would that be ok? Do you have any questions?"
3. Expose – the site for injection

Preparation

4. Collect and check you have all the required equipment - 5ml syringe, ampoule, pair of gloves, (2X) 21 G green needles, tourniquet, alcohol swab and a BNF
5. Check that the packaging of all the equipment is intact and in date
6. Wash your hands
7. Prepare and lay out the equipment required

8. Consult the patent's prescription sheet and ascertain the following -

9. Drug to be administered

10. Dose

11. Date and time of administration

12. Dilute as appropriate

13. Route of administration

14. Validity of the prescription

15. Ensure patient has no relevant drug allergies

Procedure

16. Confirm the name of the drug on the ampoule that you are administering. Confirm the strength and the expiry date with a nurse or doctor

17. Tap the neck of the ampoule gently. Cover the neck with a sterile swab and snap open. Ensure there are no glass fragments within the ampoule

18. Draw up the required amount of drug

19. With the needle pointing upwards, tap the syringe and squeeze out any air whilst ensuring no drug is lost

20. Discard the needle in the sharps bin and attach a new green needle

21. Position the arm ready for venepuncture and identify a suitable vein. Apply the tourniquet

22. Use an alcohol swab to clean the area to be injected and allow to dry

23. Retract the skin with your non-dominant hand and inform the patient of a sharp scratch before the needle pierces the skin

24. Observe for flashback in the needle

25. Loosen the tourniquet and then administer the drug at the correct speed and rate

26. Withdraw the needle and apply compression to the site of the injection using cotton wool to help prevent a haematoma

27. Record the time the drug was administered on the drug chart and sign

29. Dispose of your sharps in the sharps bin and place the waste in the clinical waste bin

33. Check the patient is comfortable

34. Thank the patient

35. Wash your hands

Questions that you could be asked

- What are the possible complications of administering an injection?
- Can you name three drugs that can be given by injection?

Catheterisation

Things to remember

- Because of the personal nature of this station, preparation of the equipment should be completed prior to exposure of the patient
- It is often a good idea to put on two pairs of gloves at the start so that there is no need to wash your hands and put on a second pair of gloves halfway through
- Never touch the penis directly with your glove - always use a swab

ICE

1. Introduce - offer a chaperone for the procedure
2. Consent - explain the procedure and obtain consent. "I have been asked to place a flexible tube through your penis into your bladder, it may feel a little uncomfortable but it shouldn't be painful. Would that be ok and do you have any questions before I start?"
3. Expose – ask the patient if they can remove their trousers and underwear and lay down on the bed with their legs parted. Inform them that you will just step outside to prepare the equipment and wash your hands

Preparation (dirty)

4. Ensure an aseptic technique - put on an apron and clean the trolley
5. Wash your hands

6. Prepare a sterile field. Peel the outer plastic covering of the catheterisation pack and slide the pack onto the trolley. Open without touching the inside of the pack to form a sterile field in which to work

7. Put on a pair of sterile gloves

9. Stick the yellow disposable bag on to the side of the trolley. Pour sterile water (for cleaning) into a small bowl with swabs

10. Ensure the patient is exposed adequately and retract the foreskin

11. Place the drape around the penis and hold the penis with a sterile swab in your non-dominant hand

13. Clean around the sides of the penis and the urethral meatus with swabs that are soaked in saline

16. Remove gloves, dispose of them and ask for a second pair of sterile gloves (if not wearing them underneath)

Catheterisation (clean)

17. Wash your hands and put on second pair of gloves

18. Hold the shaft of the penis with a sterile gloved hand and swab. Insert the lignocaine gel by squeezing a small amount (5m1) into the urethra while the penis is held vertically

19. Allow 5 minutes for the anaesthetic to work whilst holding the penis vertically

20. Place the catheter, still in the inner plastic covering, into the receptacle and put it between the patient's legs

22. Dip the tip of the catheter into the lignocaine jelly

23. Hold the penis vertically and insert the catheter into the urethra holding only the plastic covering. If resistance is felt, ask the patient to cough or relax as if they were going to pass urine. Insert the full length of the catheter

24. Inflate the catheter balloon with 1 ml of sterile water and ask the patient if this causes pain or discomfort. If none is experienced inject the remaining 9 ml of sterile water. Pull back gently on the catheter

25. Attach the drainage bag to the end of the catheter and reposition the foreskin (avoiding a paraphimosis)

26. Dispose of the waste appropriately in a clinical waste bin

27. Document in the notes the size of the catheter used and the residual volume of urine initially collected

28. Thank the patient

29. Offer to cover them up and check they are comfortable

30. Wash your hands

Questions that you could be asked

- What are the main indications for catheterisation?
- What are the risks involved in catheterisation?

Rectal examination (on model)

Things to remember

- Request a chaperone for this procedure
- Findings to comment on when palpating the prostate gland - size - normal or enlarged, shape - regular or irregular, surface - smooth or uneven, consistency - firm, rubbery or hard, central sulcus - present or absent, rectal mucosa - mobile or fixed

ICE

1. Introduce
2. Consent - explain to the patient that "because of the symptoms that you have been experiencing I would like to examine your back passage using my finger to see if there are any abnormalities there. Would that be ok? The examination may feel a little uncomfortable but should not be painful"
3. Expose - ask the patient if they can remove their underwear and lay on their left hand side with their buttocks at the edge of the bed and knees drawn up towards their chin (foetal position). Inform the patient you are going to step outside to prepare the equipment and wash your hands

Preparation

4. Request the presence of a chaperone and inform the patient of this
5. Wash your hands and put on a pair of gloves

Inspection

6. Gently part the buttocks and inspect the anus and the surrounding skin. Look for scars, excoriations, skin tags, ulcers, fissures, polyps, mucosal prolapse or external haemorrhoids

Palpation

7. Lubricate the gloved index finger of your right hand
8. Warn the patient that you are now going to exam the back passage
9. Feel for induration of the perianal skin
10. Insert your finger as if pointing towards the genitalia
11. Assess the anal sphincter tone by asking the patient to tense and squeeze on your finger
12. Palpate the entire circumference of the rectum by rotating your hand clockwise and anticlockwise feeling for any masses or indurated tissue
13. Palpate the rectum noting if it is loaded with stool, or if it is empty. Note the consistency of any faeces
14. Palpate the prostate - in males palpate the prostate gland noting its size, shape, surface, consistency, tenderness and the presence of a midline groove
15. Remove your index finger and examine the stool found on the glove. Look in particular for the stool colour and the presence of blood or mucous
16. Clean - wipe off any lubricant remaining on the anus and remove any faeces on the anal margin using a gauze or tissue
17. Inform the patient that the examination is complete and check they are not experiencing any discomfort

18. Give the patient some tissue and tell the patient that you are going to step outside whilst they get dressed. Offer to assist them with this (cover the model)

19. Remove and dispose of the gloves along with other waste in the clinical waste bin

20. Thank the patient and ask if they have any questions

21. Wash your hands

22 Record findings in the notes

Questions that you could be asked

- What investigations would you performed if a prostate mass is found on examination?
- How is benign prostatic hyperplasia treated?

Suturing

Things to remember

- Suture size and type – rule of thumb – more intricate area = bigger size sutures. Deep wounds = absorbable, superficial wounds = non absorbable
- Face – 6/0, hands – 5/0, limbs – 4/0, trunk 3/0, scalp 3/0
- Time of removal typically – face – 3-5 days, scalp – 7 days, trunk, limbs and hands – 10 days
- Absorbable sutures – monocryl, vicryl or dexon
- Non absorbable – nylon or prolene

ICE

1. Introduce
2. Consent – explain the procedure and obtain consent. Explain that you would like to place some stitches in the wound to aid healing. Reassure the patient that you will use local anaesthetic to minimise the discomfort caused
3. Expose – ensure the wound and surrounding skin are clearly visible. Ensure the comfort of the patient before proceeding

Preparation

4. Collect the required equipment – a pair of sterile gloves, syringe, 2 Needles (green and blue), 1% lignocaine vial, suture pack, antiseptic solution and swabs, suture needle holder, forceps (one toothed and one normal), scissors and a sharps bin
5. Wash your hands

6. Put on your gloves

7. Inspect the wound for any debris and dirt. This would require cleaning and debridement before continuing with the procedure

8. Open the suture pack using a sterile technique and drop a pair of sterile gloves, syringe, sutures and both needles into the field

9. Clean the wound from the centre outwards with antiseptic soaked swabs and then drape the field

10. State what local anaesthetic is to be used - type and strength - select the syringe and attach a 21 G (green) needle and draw up 5ml of 1% lignocaine - dispose of the needle and attach a 25G (blue) needle

11. Inject blebs of lignocaine into the skin encompassing the wound approximately 0.5-1cm from its edge. Aspirate the needle on inserting into the skin to ensure a vessel has not been entered. Dispose of the needles in the sharps bin and tell the examiner you would wait up to 10 minutes to allow sufficient time for the anaesthetic to work

12. Select the suture material required

Procedure

13. Gather the needle holder, scissors and forceps

14. Hold the suture needle 1/3 of the way along its length

15. Begin at the middle of the wound about 5mm from the edge. Elevate the skin with toothed forceps and pass the needle through until it emerges from inside the wound

16. Re-insert into the wound and pass under the opposite wound edge. Ensure the exit point is again 5mm from the wound edge on the opposite side

17. Pull the suture material through leaving approximately 5cm of material at the initial entry site

18. Perform a surgical knot - hold the needle end with the forceps and wind two loops of the long end of the suture in a clockwise motion around the needle holder. Pull the short end of the stitch through the loops using the mouth of the needle holder

20. Pull the knot gently down to the skin, opposing the two sides

21. To complete the knot, repeat the whole process but this time, wind once or twice anticlockwise and pull the end of the stitch through. The suture material may then be cut to complete the stitch. Each stitch should be spaced 5 - 10mm apart

22. Offer to place the next suture (using the rule of halves) half way between the present suture and the distal/medial end of the wound

23. Say you would dress the wound appropriately once suturing is completed

Closing up

24. Dispose of sharps in the sharps bin and dispose of gloves and clinical waste in the clinical waste bin

25. Thank the patient and check they are comfortable

26. Make sure you document the procedure in the notes along with the suture type used. You should also state if the sutures are to be removed and when this should take place

27. Wash your hands

Questions that you could be asked

- When would you consider matrice sutures and why?
- What are the complications of suturing?

Phlebotomy

Things to remember

- Know the common bottle types for the various tests. This can vary in different hospitals but typically purple is for haematology, yellow for biochemistry, blue for clotting and pink for group and save/crossmatch

ICE

1. Introduce – check patient age, date of birth and the first line of their address
2. Consent – explain the procedure to the patient and ask for their consent
3. Expose - correctly position the patient with their preferred arm horizontal and fully extended

Preparation

4. Collect the required equipment - non sterile gloves, vacutainer needle, vacutainer holder, blood bottles, cotton wool, alcohol swab, tape, and tourniquet
5. Wash your hands
6. Put on non sterile gloves

Procedure

7. Apply tourniquet to the upper arm above the antecubital fossa

8. Select vein - choose an appropriate vein by palpation - mention techniques that may help reveal a vein such as gentle tapping, warming the skin or making and releasing a fist

9. Clean the patient's skin carefully using an alcohol wipe and allow time to dry

10. Retract the skin a few centimetres below the proposed insertion site to stabilise the vein

11. Insert the needle smoothly at an angle of approximately 15 – 30°

12. Level off the needle once puncture of the vein wall is felt and advance the needle by a further 1mm

13. Hold the needle steady and attach the blood bottles to the vacutainer system

14. Release the tourniquet and place some cotton wool over the puncture site once the needle is removed. Apply pressure for 1 minute ensuring that the bleeding has stopped

15. Ensure adequate mixing of the tubes if they contain anticoagulant

16. Label the bottles with relevant details (name, date of birth, address, hospital number and date and time of blood collection)

17. Inspect the puncture site before applying a dressing ensuring they are not allergic to plasters if this is to be applied

18. Discard sharps into the sharps bin and the remaining waste into the clinical waste bin

25. Thank the patient and ensure they are comfortable

26. Offer to complete a blood request form and complete all relevant sections, including clinical details as described above

27. Wash your hands

Questions that you could be asked

- What are the risks of phlebotomy?
- What should be done if a needle stick injury occurs?

ECG - Placing of electrodes and interpretation of an ECG

Things to remember

- ECG's should always be analysed in a step by step format. Don't jump to the obvious abnormality as you may miss many other important findings
- Know the position of the chest leads - V1 = 4th intercostal space at the right sternal edge, V2 = 4th intercostal space at the left sternal edge, V3 = midway between V2 and V4 , V4 = 5th intercostal space in the mid clavicular line, V5 = 5th intercostal space, anterior axillary line, V6 = 5th intercostal space, mid axillary line
- In a female patient. V4, V5 and V6 are placed underneath the left breast
- The limb leads should be attached in a clockwise fashion in the order of traffic lights (red - right arm, yellow - left arm, left leg - green and right leg - black)

Recording an electrocardiogram

ICE

1. Introduce – introduce yourself and elicit the name, date of birth and first line of the patient's address
2. Consent - explain the procedure to the patient. Reassure the patient that the procedure will not cause any pain
3. Expose - lay the patient on the couch and expose the patient's arms, ankles and chest

Procedure

4. Ensure the ECG machine is in working order. Ensure the correct date, time and settings are programmed and that suitable paper is available

6. Inspect the skin to determine if skin preparation is required - e.g. removal of chest hair or cleaning of the skin

7. Attach the limb leads to the dorsal aspect of the forearms and on the outer aspect of the lower limbs above the ankles. Ensure good contact between the electrode sticky pads with their adjacent leads

8. Place chest leads in the correct intercostal spaces and attach them correctly to the ECG machine

9. Turn the ECG machine on. Check the calibration and print speed (25mm/sec) and press 'filter' and then 'start' to print the ECG

10. Recheck the patent's details for identification and write down the patient's name, date of birth, address, hospital number and the time and date when the ECG was taken on the ECG

11. Identify heart rate from the ECG (300 divided by the number of large squares between adjacent R waves)

12. Assess the cardiac rhythm (check that the P wave is followed by a QRS complex)

13. Determine if the axis of the ECG is normal

14. Assess the morphology of the P wave for abnormalities

15. Measure the PR interval (or state if absent)

16. Assess the morphology of QRS complex for abnormalities

17. Assess the ST segment for abnormalities

18. Assess the T waves for abnormalities

19. Remove the leads and dispose of any waste in the clinical waste bin

20. Thank the patient and check they are comfortable

21. Wash your hands

Questions that may be asked

- How would you analyse an ECG in a step by step approach?
- What ECG changes are used in the criteria for thrombolysis?
- Which leads look at the inferior part of the heart?

History Taking

History taking, whilst sometimes boring, is the single most important skill that you can acquire at medical school. History taking is easy honestly! With a few simple techniques applied every time you will be a history taking connoisseur in no time at all.

This area of OSCE practice could have a book written about it and there are many books that have practice scenarios. In this book I want to provide you with some essential tips on structure that I have picked up along the way.

Important tips

- Introduce yourself and confirm who they are – verify the information on why they have come to see you

- Try to formulate both a differential diagnosis and a problem list

- Always ask if there is anything you might have missed at the end of the consultation, the actors WILL try and help

- Summarise as you go, this will help remind you what you have asked and where you are going next – use a summary if you have a memory blank to help remind you where you were

- The history of presenting complaint is where all the money is! By the end of the history of presenting complaint you should have a fair idea of what is going on

- Before you start taking the history in the OSCE... indeed *before* you start taking the history in real life, take the presenting complaint... i.e. chest pain and write out a list of differential diagnosis of chest pain. You should then use the history of presenting complaint to ask questions about these differentials to determine which one it is

- The second important thing you should do with your history of presenting complaint, and I still do this with every patient I see today in medical admissions, is write four headings under the HPC title. These should be used to help remind you about questions you need to ask and will help when you present to the examiner/consultant. The four headings are:

- 1) Socrates – which you will all be familiar with. 2) Associated factors – which should include the symptoms from the relevant system (see below) 3) Risk Factors 4) Previous investigation and management. The fourth of these is very rarely asked... but should be!

- In the remaining part of the patient's history you should aim to cover all the outstanding parts of the normal history structure - past medical history/past surgical history, drug history – including over the counter medications and allergies, family history, social history and systems review

At the end of the history taking be prepared to:

- Give a one line summary of the case
- Give a differential diagnosis - (however you have already done the hard work at the start – now just put the most likely one at the top)
- Say you would want to perform some basic examinations and perform a few simple investigations - be logical

So when considering history taking you should use six main headings to guide you:

1) Differential diagnosis
2) History presenting complaint – including the four main headings
3) Cover all remaining headings of your traditional history structure
4) Summary
5) Differential diagnosis – based on history
6) Examinations and investigations

Below is the common scenario of chest pain with the different aspects of the history taking process highlighted

Practice scenario 1: Chest Pain

Please take a history from this 50 year old gentleman who presents with sudden onset chest pain

1) Differential Diagnosis

- Cardiac – e.g. Angina, MI, tamponade, pericarditis, ruptured aortic aneurysm, valve disease
- Respiratory - e.g. Pneumothorax, pulmonary embolism, pleurisy
- Gastrointestinal – e.g. acid reflux, pancreatitis, peptic ulcer, anaemia
- Musculoskeletal – e.g. muscle strain, costrochondritis

2) History presenting complaint – including four main headings

- SOCRATES - site, onset, character, radiation, associated factors (separate heading underneath), timing, exacerbating/relieving factors, severity

- Associated factors – this should include all the systems included in your differential. In the example above this would be cardiovascular - (orthopnoea, PND, ankle oedema, syncope, palpitations etc) Respiratory – (wheeze, cough, SOB, haemoptysis etc) gastrointestinal (dysphagia, odynophagia, acid reflux etc)

- Risk Factors e.g. cardiac risk factors (FCHADS - family history, cholesterol, hypertension, age, diabetes, stroke/smoking). PE risk factors (calf pain, recent travel, immobility, FH, malignancy)

- Previous investigation and management- for example if someone has chest pain it would likely be very useful to know if they have had an angiogram last week! Take it from someone who has learnt the hard way on consultant ward rounds having not asked

3) Cover all remaining headings of your traditional history structure (it's just like your driving test - be seen to be covering everything)

- Past medical history/past surgical history - angina, COPD, PE, dyspepsia, cardiac operations, pacemaker, prosthetic heart valves
- Drug history – including over the counter medications and allergies – aspirin, warfarin, GTN, antihypertensives
- Family history - are their parents alive and well? IHD, congenital heart disease
- Social history – smoking history, alcohol, recreational (coronary vasospasm – cocaine), exercise capacity, occupation, effect on life
- Systems review- Briefly cover all systems

4) Summary

In summary he is a 50 year old smoker who presents with sudden onset chest pain on the background history of ischaemic heart disease.

5) Differential diagnosis - based on history – they will want to hear the most common and serious causes of chest pain which you would not want to miss on day 1 as a junior doctor

- Myocardial infarction
- Aortic dissection
- Angina
- Pneumothorax
- Pulmonary embolism
- Pneumonia (pleuritis)

6) Examinations and investigations - there are loads so just mention the most common ones.

Observation chart, cardiovascular examination (including peripheral pulses), respiratory examination

Blood Tests - FBC (Hb - anaemia wcc -infection), U and E's, CRP, D dimer, Troponin initial and 12 hour
ECG, CXR, Consider CT Aorta/CTPA, ECHO

Practice scenario 2: Change in bowel habit

Please assess this 42 year old gentleman who presents with a 4 week history of change in bowel habit

1) Differential diagnosis

- Colon cancer
- Diverticular disease
- Inflammatory bowel disease
- Ischemic bowel
- Irritable bowel
- Diet
- Medication

2) History presenting complaint – including four main headings

- SOCRATES - onset (sudden/chronic) character (frequency/colour), timing (how long), associated factors (separate heading underneath), exacerbating/relieving factors, (medications, travel, diet), severity (how many motions a day, effect on life)

- Associated factors – diarrhoea, constipation, PR blood, vomiting, haematemesis, abdominal pain, fever, tenesmus, mucous, jaundice, anaemia

- Risk factors -travel, change in diet, previous malignancy, atrial fibrillation

- Previous investigation and management- colonoscopy, sigmoidoscopy, OGD

3) Cover all remaining headings of your traditional history structure

- Past medical history/past surgical history - IBD, malignancy, abdominal surgery
- Drug history – laxatives
- Family history - are their parents alive and well? IBD, malignancy
- Social history - smoking, alcohol, travel history, occupation
- Systems review- briefly cover all systems

4) Summary

In summary this is a 42 year old gentleman who presents with a 4 week history of worsening bloody diarrhoea

5) Differential diagnosis – based on history (they will want to hear the most common and serious causes of change in bowel habit which you would not want to miss on day 1 as a junior doctor)

- Carcinoma colon
- Diverticular disease
- Inflammatory bowel disease

6) Examinations and investigations (there are loads so just mention the most common ones)

Observations, abdominal examination

Blood Tests: FBC (hb – anaemia, wcc – infection), haematinics ESR (inflammation, malignancy), CRP, LFT's, calcium (metastases), TFT's

Microbiology: Stool MC+S + C Difficile toxin

Imaging: sigmoidoscopy, colonoscopy, Barium follow through

Practice scenario 3: Cough

Please take a detailed history from this 53 year old female who presents with worsening cough

1) Differential diagnosis (there are many!)

- Infection
- PE
- Asthma
- COPD
- Lung cancer
- Drugs
- GORD
- CCF

2) History presenting complaint – including four main headings

- SOCRATES – onset – acute/chronic, character – worse at night (asthma), exacerbating (occupation, pets), relieving (avoidance allergens, inhalers) severity (progression)
- Associated factors – sputum, haemoptysis, wheeze, SOB, pleuritic pain, fever, orthopnoea, PND, ankle swelling
- Risk factors: - PE (travel, immobility, DVT) acid reflux, industrial exposure
- Previous investigation and management - CXR, bronchoscopy, ECHO

3) Cover all remaining headings of your traditional history structure

- Past medical history/past surgical history - asthma, COPD previous MI
- Drug history – ACE inhibitors
- Family history - malignancy

- Social history - smoking, alcohol, travel history, occupation, pets, exercise tolerance
- Systems review - briefly cover all systems

4) Summary

In summary this patient is a 53 year old female presenting with a 2 week history of cough

5) Differential Diagnoses – based on history (they will want to hear the most common and serious causes of cough which you would not want to miss on day 1 as a junior doctor)

- Infection
- COPD
- Asthma
- PE
- Lung cancer

6) Examinations and investigations

Observation chart, respiratory examination, cardiovascular examination

Blood Tests: FBC (Hb – anaemia wcc – infection), ESR (inflammation, malignancy), CRP, D dimer ABG
Pulmonary function tests
Radiology: CXR, CTPA,

I hope from the examples above that you now have a fair idea on how best to approach the history taking stations. Spend some time practicing and preparing some history plans for the scenarios below.

Practice scenario 4: Collapse

Please take a history from this 57 year old female who presents with recurrent collapse episodes

Practice scenario 5: Acute abdominal pain

This 18 year old female presents with sudden onset abdominal pain. Please take a history and provide a differential diagnosis

Practice scenario 6: Back pain

Please take a history from this 47 year old gentleman who has been referred by his GP with back pain

Practice scenario 7: Falls

Please take a history from this elderly 90 year old patient who presents with recurrent falls

Practice scenario 8: Confusion

This lady has presented to the assessment unit acutely confused, please take a history to determine the cause

Practice scenario 9: Haematemesis

Please could you kindly take a history from this 38 year old male who has presented with haematemesis

Practice scenario 10: Joint Pain

Please take a history from this 55 year old lady who presents with joint pain

Acute Management

The acute management stations are supposed to replicate "real life" scenarios. As anyone who has read the "real" junior doctor survival guide will be able to testify, general practitioners are not always my favourite people... but actually they are very good at performing focussed history and examinations in a limited time... the only difference is your management plan needs to be more comprehensive than simply "refer to secondary care!" Up until now I have told you to use ICE at the start of your examinations (ICE - introduce, consent, expose), now I want you to use ICE at the end (ideas, concerns, expectations). Ideas - what the patient thinks might be wrong with them. Concerns - what they are worried about ("Doc, I think it might be cancer"). Expectations - what they want to happen from here. So from now on you can act as COOL AS ICE AT THE START AND END... cheesy, I know!

When you receive the short lead in to the station - the first thing you need to think is "what are the differentials for this complaint?" Once you have a set of differentials you simply need to ask questions and perform examinations to decide which one is more likely.... not to different from real life I hear you cry. In the history part, try to ask all sections of your normal history, HPC, PMH, DH etc, but try to do this in a way that shows the examiner that you have a differential in mind. This should be clearer once you have looked at the examples below.

So using these simple four questions below have a go at the scenarios underneath.

1. What are the essential differential diagnoses?

2. What essential questions must you ask in the history?

3. What essential examinations should you perform?

4. How would you investigate and simply manage this in the assessment unit?

Scenario 1: Leg Swelling

The GP has asked you to see this 60 year old lady who has recently been an inpatient in your hospital. She has presented to the GP practice with a swollen right calf.

Question 1 – What are your differential diagnoses?

- Deep vein thrombosis - this is likely to be top of your list and should help guide your consultation

- Ruptured Baker's cyst

- Compartment syndrome

- Cellulitis

Question 2 – What essential questions must you ask in the history?

HPC - SOCRATES (in particular onset - when did it start? Was it sudden onset), Associated factors (SOB - PE/DVT, Pain, change in skin colour, fever/sweats), Risk factors – Immobility, surgery, malignancy, history of DVT, trauma (compartment syndrome/cellulitiis), rheumatoid arthritis (ruptured bakers cyst). Previous investigation – previous Doppler/CTPA

PMH - use this section of your history to name your differentials, "any past history of DVT, Bakers cyst" - tells the examiner what you're thinking

DH - anticoagulants

FH - clotting disorders

SH - Who is at home with them? Ideas, concerns and expectations (ICE)

Question 3 - What essential examinations should you perform?

As a good rule of thumb always look at the observation chart and listen to heart and lungs - just like you would always do in the assessment unit

Observation chart - temperature (fever - cellulitis) low oxygen saturations (PE)

Examine lower limb - skin changes, measuring size of calves, peripheral pulses/sensation - compartment syndrome

Examine the chest - pleural rub - PE

Question 4 - How would you investigate and simply manage this in the assessment unit?

Do the simple tests first!

Bloods - FBC, CRP, D dimer, CK

ECG, CXR, doppler

Think how you would manage the most likely diagnosis – in this example DVT – anticoagulation and analgesia

Scenario 2: Transient visual disturbance

The GP has asked you to see this 65 year old lady who has presented with an episode of transient loss of vision with full recovery

Question 1 – What are your differential diagnoses?

- TIA

- Retinal or vitreous haemorrhage

- Central retinal vein or branch occlusion

- Temporal arteritis

- Diabetic maculopathy

Question 2 – What essential questions must you ask in the history?

HPC – SOCRATES (in particular onset – when did it start? Was it sudden onset?), character – partial or complete, one eye/both eyes. Timing – low long did it last, complete recovery?. Associated factors - facial or limb weakness/parathesiae, speech disturbance, headache. Risk factors - cardiovascular risk factors scalp tenderness, joint pains (PMR/Temporal arteritis)

PMH – use this section of your history to name your differentials, "any past history of diabetes, stroke, atrial fibrillation, migraines or polymyalgia rheumatics/temporal arteritis" – tells the examiner what you're thinking

DH – review drug history, anticoagulants

FH – stroke/diabetes

SH – Who is at home with them? Ideas, concerns and expectations (ICE)

Question 3 - What essential examinations should you perform?

Observation chart – blood pressure (BP important for ABCD2 criteria), BM (diabetes)

Examine cardiovascular - assess pulse (rate/irregular), carotids (bruits), murmurs (embolic source)

Examine eyes – acuity, visual fields, fundoscopy, palpate over the scalp (temporal arteritis)

Examine neurological system – time is likely to be short by now! Therefore tell the examiner that you would like to perform full cranial nerve, upper limb and lower limb exam – if time allows check for facial weakness and power in the limbs

Question 4 - How would you investigate and simply manage this in the assessment unit?

Don't forget to address concerns and do the simple tests first!

Blood tests - lipid profile, fasting glucose, ESR

CT head, ECHO/24 Hour tape/carotid doppler

Management - most likely diagnosis is TIA, if AF - likely will need warfarin, optimise risk factors, TIA review (know about the ABCD2 criteria)

Scenario 3: Headache

The GP has asked you to see this 70 year old gentleman who has presented with worsening headaches.

Question 1 - What are your differential diagnoses? (clearly there are many)

- Haemorrhage

- Meningitis

- Tension Headaches

- Migraine

- Cluster headaches

- Brain Tumour

- Trigeninal Neuralgia

Question 2 - What essential questions must you ask in the history?

HPC - SOCRATES (In particular onset - when did it start? Was it sudden onset?), location, character (intermittent/persistent), radiation, exacerbating and relieving factors, timing - clusters, early morning Associated factors - facial or limb weakness/parathesiae, pains in the face, visual disturbance, fits, feints and funny turns, aura, vomiting, pain on bending forward/coughing, fever and a rash. Risk factors - known migraines, trauma, analgesia or stress

PMH - use this section of your history to name your differentials, "any past history of migraines, trigeminal neuralgia, tension headaches" - tells the examiner what you're thinking

DH - review drug history, particularly analgesia, side effects,

FH - migraines

SH - who is at home with them? Ideas concerns and expectations (ICE)

Question 3 - What essential examinations should you perform?

Observation chart - blood pressure

Examine eyes - acuity, visual fields, fundoscopy, signs of papiloedema/optic atrophy,

Examine cranial nerves: eye movements/diplopia, trigeminal neuralgia, facial nerve weakness, and palpate the temporal arteries

Examine neurological system - upper limb and lower limb exam - if time allows

Question 4 - How would you investigate and simply manage this in the assessment unit?

Don't forget to address concerns (this is likely to be "I'm worried this is a brain tumour") and do the simple tests first!

Blood Tests - FBC (wcc – infection) clotting (haemorrhage), ESR (temporal arteritis) Blood cultures

CT Head/MRI +/- LP

I Hope from the examples above that you now have a fair idea on how best to approach the acute management scenarios. The topics above probably represent the tougher end of what might be expected. You are likely to be given ones such as chest pain, SOB and diarrhoea in the actual exam, and I am hopeful that you will have had enough exposure to these in the admissions unit to apply the structure above and ace them with no problem at all. Spend some time practicing the scenarios below with your friends

Scenario 4: Chest pain

This 58 year old gentleman has presented with sudden onset central chest pain. Please review

Scenario 5: Shortness of breath

Accident and Emergency have referred this 30 year old gentleman who was admitted with sudden onset breathlessness

Scenario 6: Hemiplegia

The GP has referred this 80 year old lady with new onset left sided weakness

Scenario 7: Confusion

Please could you assess this 85 year old gentleman who has presented with new onset confusion

Scenario 8: Acute abdominal pain

The GP has referred this 19 year old female who has presented with acute abdominal pain

Scenario 9: Diarrhoea

This 28 year old patient has been referred from the emergency department with a one week history of bloody diarrhoea

Scenario 10: Collapse

Please assess this 68 year old lady who has presented with two recent collapse episodes

Communication in a difficult or challenging situation

The communication stations are often neglected somewhat in the time allocated to their practice in the run up to finals. It is often assumed that it is all common sense! Whilst this may be true in part, not all medical students are blessed with common sense and those that are, may not necessarily be gifted with the world's greatest people skills!

With this in mind, the pages below will outline a little knowledge and a step by step approach that will go a long way to impressing the examiners. I will also provide some practice scenarios for you to work through to make sure you get every last mark you can in this potentially quite straightforward part of the exam.

Things to remember

Breaking bad news

- This part of the exam attempts to find candidates who can be sensitive
- Empathy is the key
- Always identify what the patient knows
- Lay a foundation of information then build on it
- Try not to explore the history – this is not a history taking station
- Remember patients can often only remember three things from a consultation – always summarise and make a plan
- For this scenario it is likely to be breaking bad news/cancelled operation/angry relatives
- Check their understanding early
- Give warning shots

- Stay silent for 3 seconds more once you start feeling awkward – easier said than done, I know!

Dealing with an angry patient or relative

- Confirm who they are
- Acknowledge that the patient is angry or upset and let them know that you are aware of this and you can understand why
- Allow them to tell you how angry they are but never react – remember its acting!
- Try to provide some simple solutions to help overcome these issues
- If there has been a mistake, admit it and apologise for it, explain that you will escalate to your more senior colleagues and that you will follow the Trust's complaints procedure

Breaking bad news

Beginning the session

1. Introduce yourself to the patient, confirm their identity, and ask the patient their understanding of why they have come to see you
2. Ask them if anyone else has come with them today and ask if they would like anyone else to be present
3. Ask them what has happened to this point – symptoms, investigation and management
4. Ask them what their main concerns are and what they think may be going on

5. Review the investigation with them and explain that you now have all the results available

Sharing the information

6. Elicit - assess the patient's understanding first - what the patient already knows
7. Gauge how much the patient wishes to know
8. Give a warning shot first, ideally a couple of warning shots, that difficult information is coming e.g. - "I'm afraid it looks more serious than we had hoped"
9. Give basic information, simply and honestly, "unfortunately there is a tumour in the bowel, and is likely to be cancer, I am so sorry to have to tell you this"

Being sensitive to the patient

10. Pause – allow time for the patient to take the information in, wait for them to speak next
11. Gauge the patients need for further information as you go and give more information as requested. Listen to the patient's wishes as individuals vary greatly

Planning and support

12. Provide a plan – "although it may not be curable there is plenty that can be done to help. I am going to put you in contact with the cancer specialist and I am very hopeful they will be able to see you early next week"
13. Elicit what the patients main concerns are at this point and provide simple advice
14. Reiterate that there is plenty of support available to them

15. Explain that you will have them speak to a specialist nurse today and give them a point of contact

16. Explain that they will probably have loads of questions once they leave but ask if they have anything they want to ask now

Ending the consultation and providing follow up

18. Offer to speak to their family if they would like
19. Provide written information and sources of support
20. Explain that you will see them next week in clinic and offer that they can bring someone with them

Practice scenarios

Scenario 1

You are asked to see a 58 year old gentleman with the results of his CT scan. He had been admitted with persistent cough and haemoptysis by his GP. CXR showed a mass in the left upper lobe and the CT scan shows multiple metastases. Please discuss these results with the patient.

Scenario 2

You are the junior doctor on call for the weekend and you are asked to speak to the son of a 75 year old lady who has been prescribed amoxicillin for pneumonia despite it saying clearly on the drug chart that she is allergic to penicillin. He is angry and would like to know how this could have happened

Scenario 3

You are the junior member of the trauma and orthopaedic team. Mr Jones was due for a knee replacement today but due to several emergencies overnight his operation has been cancelled. Please inform the patient of this and answer any questions he may have.

Explanation station (giving information - explaining a procedure/diagnosis)

Things to remember

- Follow a set sequence regardless of the procedure/diagnosis to explain - this is not a test of knowledge it is a test of the way you can communicate and interact with patients
- Always find out what they understand about the procedure and its potential benefits
- Always ask what concerns they have and why they are worried

Explaining a procedure

Introduction

1. Introduction - introduce yourself and elicit the name and age of the patient. Try to establish a rapport
2. Understanding - elicit patients understanding of what has happened till now and why they are having the procedure done
2. Concerns – elicit any particular patient concerns about the procedure and explain that you are happy to go through these with the patient

Explaining

4. Explain procedure - clarify to the patient what the procedure is and clarify any abbreviations e.g. OGD
5. Pre-procedure - explain to the patient what will happen before the procedure

6. Procedure - explain to the patient what will happen during the procedure

7. Post-procedure - explain to the patient what will happen after the procedure

8. Benefits and risks - explain benefits and risks of the procedure outlining alternatives if available

Closing up

9. Understanding - check whether the patient had understood what has been explained

10. Questions - encourage patients to ask questions and deal with them appropriately

11. Summary - provide an appropriate summary of the procedure to the patient and reiterate why it is being proposed

12. Follow up - mention the need for outpatient follow up for results and to discuss further management plans

13. Respond - appreciate that it is a large amount of information to take in and acknowledge the patients feelings

14. Leaflets - offer patient leaflets regarding the procedure and a person of contact if they have any further questions

15. Thank the patient

Practice scenarios

Scenario 1

A 48 year old gentleman was admitted to the ward yesterday with haematemesis and has been scheduled for an OGD. Please explain this procedure to him and address any concerns that he may have

Scenario 2

A 28 year old gentleman has been admitted with recurrent episodes of diarrhoea. The diagnosis of inflammatory bowel disease has been suggested and a colonoscopy has been arranged. Please explain the procedure and answer any questions he may have

Scenario 3

A 58 year old female has been admitted with an NSTEMI and has been scheduled to undergo coronary angioplasty. Please explain the procedure to the patient and elicit any concern they may have

Explaining a diagnosis

This is very similar to explaining a procedure except that it is likely to be more straightforward as the patient is likely to already have questions that they want to ask

Introduction

1. Introduction - introduce yourself and elicit the name and age of the patient. Establish rapport
2. Understanding – ask the patient if they understand why they have been invited to attend today and what has happened up to this point
3. Review – summarise the previous investigations

Explaining

4. Explain the diagnosis - explain what the likely diagnosis is having reviewed all the results
5. Allow the patient time to take in the information given and await any questions
6. Ask if they were expecting this and what their understanding is of this diagnosis
7. Offer to give them an overview of the diagnosis – keep this simple

Closing up

8. Reassure – empathise with the patient and try to reassure them that there is plenty of support and treatment approaches available

9. Concerns – ask the patient if they have any other specific concerns or if they have any questions or worries

10. Summary – summarise the information given

11. Follow up - mention need for outpatient follow up and that you will meet next week to formulate a further plan and address any new questions or concerns

12. Leaflets - offer patient information and a person to contact if they have further questions

13. Thank the patient

I hope this section has provided you with a simple approach to the communication section of the exam. Practice the scenarios below with your friends using the structure outlined.

Practice scenarios

Scenario 1

A 55 year old lifelong smoker has been admitted with a further lower respiratory tract infection. Pulmonary function tests have confirmed a diagnosis of chronic obstructive pulmonary disease (COPD) and you have been asked to communicate this to the patient

Scenario 2

A 63 year old gentleman was admitted on the acute take with chest pain. His 12 hour troponin has come back positive and the consultant has asked you to communicate this information to the patient and discuss any worries he may have

Scenario 3

A 28 year old lady has had investigation for visual loss and new weakness in her arm. The results of these investigations have confirmed the diagnosis of multiple sclerosis. Please inform the patient of this and answer any questions she may have

Practice mark sheet

Progress chart (example)

Exam/Date	Resp 04/02/12	Resp 08/02/12				
1	tick	tick				
2	cross	tick				
3	tick	tick				
4	tick	tick				
5	cross	cross				
6	tick	tick				
7	cross	tick				

Using the above progress chart it is easy to identify which number points in the examination sequences you are forgetting e.g. Point 5 in respiratory exam – take the respiratory rate.

Exam /Date							
1							
2							
3							
4							
5							
6							
7							
8							
9							
10							
11							
12							
13							
14							
15							
16							
17							
18							
19							
20							
21							
22							
23							
24							
25							
26							
27							
28							
29							
30							
31							
32							
33							
34							
35							
36							
37							
38							
39							
40